Mike Askew is an internationally regarded expert on mathematics education, having held professorial chairs at King's College, London, and Monash University, Melbourne, Australia. Mike has also taught and researched in North America and South Africa. Mike believes that given rich, engaging and challenging problems to reason about, everyone can act as a mathematician and develop as a mathematical thinker. His research has been widely published, both for academic and popular audiences, including the highly successful *Maths for Mums and Dads* (2010), co-authored with Rob Eastaway, and *Transforming Primary Mathematics* (2012). Mike is also an accomplished magician.

D0061825

# MATHEMATICS
## ALL THAT MATTERS

*For the boys: Russell and Max*

# MATHEMATICS

Mike Askew

ALL THAT MATTERS

First published in Great Britain in 2015 by Hodder & Stoughton. An Hachette UK company.

This edition published 2015

Copyright © Mike Askew 2015

The right of Mike Askew to be identified as the Author of the Work has been asserted by him in accordance with the Copyright, Designs and Patents Act 1988.

Database right Hodder & Stoughton (makers)

*British Library Cataloguing in Publication Data:* a catalogue record for this title is available from the British Library.

Paperback ISBN 978 1 47360173 4

eBook ISBN 978 1 47360175 8

*Library of Congress Catalog Card Number:* on file.

1

The publisher has used its best endeavours to ensure that any website addresses referred to in this book are correct and active at the time of going to press. However, the publisher and the author have no responsibility for the websites and can make no guarantee that a site will remain live or that the content will remain relevant, decent or appropriate.

The publisher has made every effort to mark as such all words which it believes to be trademarks. The publisher should also like to make it clear that the presence of a word in the book, whether marked or unmarked, in no way affects its legal status as a trademark.

Every reasonable effort has been made by the publisher to trace the copyright holders of material in this book. Any errors or omissions should be notified in writing to the publisher, who will endeavour to rectify the situation for any reprints and future editions.

Typeset by Cenveo® Publisher Services.

Printed and bound in Great Britain by CPI Group (UK) Ltd., Croydon, CR0 4YY.

Hodder & Stoughton policy is to use papers that are natural, renewable and recyclable products and made from wood grown in sustainable forests. The logging and manufacturing processes are expected to conform to the environmental regulations of the country of origin.

Hodder & Stoughton Ltd
338 Euston Road
London NW1 3BH
www.hodder.co.uk

# Contents

1 We are all mathematicians                 1

2 Counting                                  9

3 Visualizing                              21

4 Generalizing                            35

5 Pattern sniffing                        51

6 Proving                                 67

7 Connecting                              85

8 Modelling                              101

9 Puzzling                               119

100 ideas                                133

Selected bibliography                    147

Index                                    149

# We are all mathematicians

*Ours is not to question why,*
*just invert and multiply.*

North American readers may be familiar with the mantra 'Ours is not to question why, just invert and multiply' for dividing one fraction by another. Even if you have not met it before, for many people it encapsulates the point at which they gave up on mathematics. Not only does life seem too short to memorize something you will probably never need to do, but division of fractions also seems to be magic: $1/2 \div 1/3$ becomes $1/2 \times 3/1$ or $1\frac{1}{2}$. What is going on here? How can dividing $1/2$ of anything by $1/3$ result in $1\frac{1}{2}$ – an answer that looks larger than the amount you started with? If that were possible, then we could all be as rich as Midas: get half an ounce of gold, divide it by one-third and you'll have one and a half ounces of gold; keep going and soon there would be enough gold to retire on. (There is a sensible interpretation of this division, which I look at later.) No wonder mathematics can seem so mysterious.

There is so much more to mathematics than unquestioning acceptance of pre-formed facts and procedures. In this book I explore what it means both to think with mathematics and to think about mathematics. In doing so I am not trying to set out a 'mathematics primer'– there is not the space here for that and there are plenty of those on the market (see 100 Ideas). What I hope to do is lift the veil from the face of mathematics and reveal something of her beauty. I argue that mathematics is really based in common sense and that, far from being an esoteric subject accessible only to a chosen few, mathematics is better thought of as, in Cuoco, Goldenberg and Mark's term, a set of 'habits of mind', habits that, with a little help and a bit of perseverance, more people can develop and so come to appreciate what drives mathematicians. And perhaps find some enjoyment in the subject.

# ▶ Is mathematics inaccessible?

Mathematics often gets a bad press. Describing someone as 'calculating' or 'rational' is hardly flattering and mathematicians in movies or novels are often portrayed as social misfits who rarely get the guy or girl (more often the latter, as mathematics is presented as a predominantly male interest). No wonder common refrains such as 'Oh I don't care for mathematics' or 'I was never any good at it' are often said with a wistful sense of pride.

Yet professional mathematicians talk differently. They describe looking for elegant solutions to problems, of pleasure in playing around with mathematical ideas and of creating with mathematics. As the Russian mathematician Sofia Kovalevskaya said, 'It is impossible to be a mathematician without being a poet in soul.'

So why is there such a gap between the popular and professional views of mathematics? Part of the problem lies in how mathematics is often taught in schools. Mathematics is served up there as a series of decontextualized, abstract ideas, wrested from the human struggles and interactions that gave birth to those ideas. School mathematics wrests mathematics from its humanity.

I work a lot with teachers, mainly primary (elementary), many of whom are not particularly fond of mathematics and as such are typical of many people. Sometimes their dislike – fear even – of mathematics is a result of bad experiences at school, but I believe that many people never 'get inside' the mathematical mindset. A group of teachers I once worked with had come a long way on a journey of demystifying and one day I handed out scientific calculators. A teacher called me over, curious as to why her cheap supermarket calculator had a button for $1/x$ but this was missing from the more expensive models. Pointing out that the $x^{-1}$ button was the equivalent button on the other models, I took the opportunity to talk about why mathematicians define $x^{-1}$ as equal to $1/x$ (more on this later). As I completed my explanation the teachers were muttering so I asked if there was something they had not followed. 'No', one replied, 'we follow the logic of the argument, but we were just saying, "Why would anyone want to do that?"'

That sums up the mystique of mathematics. While we might not all aspire to being poets or painters, we would not question what drives some people to create poems or paintings; we have a sense, without necessarily being part of the group of artists ourselves, that artists are interested in challenging taken-for-granted perceptions and in finding new ways of seeing the world. Mathematicians are no different. The problem is that school mathematics rarely addresses the creative aspects of mathematics. School mathematics focuses on techniques, skills and procedures. Mathematicians focus on form, structure and relations. School mathematics values having a good memory. Mathematicians develop habits of mind like sniffing out patterns, looking for generalities, pushing back boundaries. These are not arcane processes or nerdy habits; they are simply an extension of our natural curiosity. Young children are often fascinated by numbers and will, without prompting, see patterns in the system and make sensible generalizations. A friend's five-year-old daughter, for example, was delighted that she had learned to count to 99, and having got a sense of the pattern, continued it in her own way; 97, 98, 99, tenty, tenty-one, tenty-two. That's a mathematical habit of mind at work – if you have a system, why stop at 99, why not carry on? The reasoning processes at play for the mathematician are no different from those demonstrated by the five-year-old; it is just that the mathematician has more things in their mathematics kitbag to play with.

# ▶ Mathematics as embodied

An argument running through this book is that mathematics is an extension of common sense, in the literal meaning of the senses that are common to human kind. The writers Lakoff and Núñez provide a persuasive argument against what they call the myth of mathematics – that mathematics is 'out there' in the structure of the world and that mathematicians have been privy to discovering eternal truths. They argue instead that mathematics is an extension of our bodily experiences, our 'common sense' of how we move in the world, how we position ourselves in the world and our experiences of acting on objects in the world. For example, the idea of placing numbers along a line often recurs in mathematics and, Lakoff and Núñez argue, this is grounded in our everyday experiences of moving along paths and the sense of order arising from places that we visit along the way.

While some hold tightly to the position that mathematics exists independently of our knowledge, the attraction of these embodied views of the discipline is that they suggest that mathematics can be more accessible to more people, that if we trace mathematical understandings back to these everyday experiences, the structure and patterning of mathematics could be more widely appreciated.

# ▶ Mathematics is hard-wired

The evidence that we have a sense of number 'hard-wired' into us also suggests that we are more predisposed to being mathematicians than we might think. Infants as young as six months show signs of surprise (measured by tracking eye movement) when a situation is set up where, for instance, two oranges are hidden, one is removed and yet subsequently two oranges are revealed. Yet they show no signs of surprise when two oranges turn into two rubber ducks, suggesting that it is the incorrectness of quantity that grabs their attention.

Even animals have been demonstrated to be able to track changes in small quantities. Birds, for example, have been demonstrated to show signs of awareness when the number of eggs in a nest changes. Sadly, none have lived up to the expectations set by the calculating horse – a famous Parisian music-hall act at the turn of the 20th century that could 'paw out' the answers to calculations put up on a chalkboard by audience members. Eventually scientists found that the horse's trainer was using his breath to subtly signal to the horse when to stop moving. The irony is that the trainer wasn't aware he was doing this and thought his horse had real arithmetical powers.

# ▶ Is mathematics discovered or invented?

A major debate in the philosophy of mathematics is whether or not the discipline is discovered or invented. The argument for it having an independent existence usually rests on the number of achievements that mathematics has enabled humankind to accomplish, but that's a bit like claiming that a cake made using a food processor provides evidence that the food processor must have pre-existed its invention. The fact that models of the world that mathematicians create have good predictive power of how the world behaves is not a sufficient argument to convince some that mathematics must already exist out there in the world, independent of human activity and thought.

It is a debate that I doubt will ever be fully resolved, partly because I agree with philosophers taking the position that all knowledge has to be a product of human consciousness. We can never know what exists independently of the bodily means through which we come to know. This is not solipsistic; I am not suggesting that all knowledge is simply 'made up'. Rather, the argument is that mathematics may well exist independently of human minds, but we can never know. There is also a pragmatic reason for siding with the argument that mathematics is a human creation – it should then be more accessible to the members of the species that invented it.

My hope is that this book, through looking at some of the processes that mathematicians engage in and some of the ideas and problems that matter to mathematicians, will enable the reader, if not to fall in love with mathematics, then at least to come to understand its nature a little better, and perhaps care a little more for it.

# Counting

*God created the integers,*
*the rest is the work of man.*

Kronecker

# ▶ The natural numbers

The history of the **integers** – the positive and negative whole numbers together with zero – reveals that it took mathematicians a long time to accept the ideas of zero and negative quantities, raising doubts that Kronecker was correct. It does seem reasonable, however, to assume that the mathematics of counting – one, two, three ... – does have its origins in the physical world.

The evidence for our earliest number words suggests that they were used as adjectives, describing properties of collections of things – two eyes, two legs, two lions. Or, again adjectively, number words placed persons or objects in order – the first person of the tribe, the second-born child, the third antelope caught. At some point, lost in time, numbers ceased to be adjectives and became treated as objects in their own right, objects in the sense of things that, while not having a concrete existence, could be thought about, much as we can think about unicorns without them existing.

The numbers that we first meet are the familiar whole numbers – 1, 2, 3 ... – so commonplace that they seem natural, which is what mathematicians called them. Our early experiences of the **natural numbers** resonate with that history of numbers as adjectives. Counting stairs as you climb them involves the **ordinal** use of the sequence of natural numbers. Counting that you have three teddies is the **cardinal** aspect.

Adults have no difficulty in coordinating and switching between the ordinal and the cardinal, but it takes years for a young child to become aware that having labelled each biscuit on a plate as one, two, three, four, five (essentially, ordering the biscuits to distinguish the counted from the yet-to-be counted), that the pronouncement of 'five' simultaneously not only labels the last biscuit in the count with the ordinal number five, but also expresses the total cardinal number of biscuits. Such lack of awareness becomes apparent when, having counted five biscuits, you ask a three-year-old to give you four biscuits and, rather than pick up a total of four, they hand over the fourth biscuit. Although, as mentioned in Chapter 1, there is evidence that the cardinality of collections of up to three items is 'hard-wired' into us, it is a step up from that to understanding how to work flexibly with the system of natural numbers.

Simple counting reveals that the distinction between whether the mathematics is in the world or whether we bring mathematics to the world is blurred. From all the pebbles on the beach, separating out a collection of five means bringing the sensibility that we can take what is randomly organized in the natural world and use the mathematics of counting to bring order – counting out five is a subtle blend of interaction between what is out there in the world and human intelligence bringing order to that world.

# ▶ Operating on the natural numbers

Mathematics begins in earnest when we start to operate on the natural numbers. Our experiences of putting two collections of objects together (e.g. taking the biscuits from two plates and putting them onto one) or removing objects from a collection (e.g. sneaking a biscuit away when no one is looking) become abstracted into **addition** and **subtraction**. Experiences such as putting three biscuits on each of four plates lays the foundation for **multiplication**, and sharing biscuits out equally is one of the roots of **division**.

Such seemingly simple arithmetic contains the seeds of much of higher mathematics. To begin to explore that, we need to look beyond the specifics of calculations like 4 + 5 or 7 × 3 and examine the underlying structures that govern the functioning of the natural numbers and operations on them.

# The identity element

Many of us learn the multiplication tables, the times tables, through chanting, with memory helped through a common first line structure to each table: one times two is two, one times three is three ... one times ten is ten. We hardly pause to think about this, but it is an example of an important mathematical idea, the **identity element**. Multiply any number by one and the number is unchanged. While one is the identity element for multiplication, there is no equivalent natural number that acts as an identity element for addition. We take it for granted that zero serves as the identity element for addition, but zero is not regarded as a 'natural' number. The natural numbers represent what is there in the world, while zero marks an absence. Does zero count (pun intended)? As we see below, the behaviour of zero makes it questionable whether it should be treated as a number at all.

## Why 'identity element' and not 'identity number'?

Identity element expresses an idea, rather than a specific number. The idea that there is a member of a set that, when combined in some way with another member of that set, leaves everything unchanged enables mathematicians to make connections across seeming disparate realms of inquiry. For example, as I explore in Chapter 7, an equilateral triangle rotated clockwise through 120 degrees or 240 degrees looks exactly the same as it did before the rotation (assuming it is plain and unmarked). A rotation of 360 degrees has the effect of bringing the triangle back to its original orientation – this rotation acts as an identity element.

# Place value

Prior to the introduction of the decimal place value system – recording all numbers with only nine digits plus zero – arithmeticians carrying out calculations involving the natural numbers had no need for zero. The Romans, for example, would have written the number 205 as CCV (the Babylonian, Sumerian and other early mathematicians had different symbols, but their systems were similar to the Roman system of having to introduce new symbols as numbers got larger). In calculating, Roman arithmeticians would not – could not – have found the answer by manipulating symbols: CCV + MCLIX cannot be answered by lining up the letters in columns. Instead, numbers were represented on a counting board, an abacus – a clay tablet with grooves in it. Small clay pellets were placed in the grooves and calculations were carried out by combining and exchanging the pellets.

In this system, 205 recorded as either CCV or CC V is unambiguous. Replacing Roman symbols with the Arabic numerals, does 25 mean the same as 2 5? Is the gap between the 2 and the 5 significant, there because the 2 is representing two beads in the hundreds column of the counting board, or is it the result of sloppy scribing? Zero steps in to 'hold the place' so that 25 is clearly different from 205. Zero does something, rather than represents something, as the other two digits do.

# Is zero a number?

Having come into being as a placeholder, with the advent of mass-produced paper, arithmetic moved away from the practical art of the abacus to paper-and-pencil algorithms. Adding two numbers column by column, zero could then be thought of as the identity element for addition (and subtraction: $n - 0 = n$). How does zero behave with the other arithmetical operations of multiplication and division? How do we make sense of calculations like $4 \times 0$, $0 \times 4$, $0 \div 4$ or $4 \div 0$?

Since, for example, $3 \times 4 = 4 \times 3$, any answer to $4 \times 0$ has to fit with that given to $0 \times 4$. Reading $4 \times 0$ as 'four multiplied by zero', it makes sense to interpret this as four taken no times, that is, nothing, zero. And 'zero multiplied by four' – $0 \times 4$ – as nothing taken four times, which again is sensibly interpreted as zero. So $4 \times 0 = 0 \times 4 = 0$, or more generally, $n \times 0 = 0 \times n = 0$ for any number ($n$).

How about $0 \div 4$? If I have zero amount of chocolate to divide out between four friends, everyone is going to get the same amount of chocolate – that is, no chocolate. Interpreting $0 \div 4$ as equal to zero would seem to keep the mathematics sweet, if not the friends. So $0 \div n = 0$ for any number ($n$).

That leaves $4 \div 0$. One way to interpret this is to say, 'suppose I have four bars of chocolate to share between zero people', then it makes sense to say 'each of my non-existent people gets no chocolate'. The answer is zero.

A different interpretation of $4 \div 0$ uses the inverse relationship between multiplication and division. Given $12 \div 3 = 4$, from the answer 4 we get back to 12 by applying the inverse operation, i.e. multiplying by three, $4 \times 3 = 12$. Multiplying by three is the **inverse** to dividing by three, and vice versa. To keep the mathematics consistent, the inverse of dividing by zero must be to multiply by zero. So suppose the answer to $4 \div 0$ is $p$. If $4 \div 0 = p$, then applying the inverse operation of multiplying by 0 we get $4 = p \times 0 = 0$. Concluding that four equals zero is, however, clearly nonsense.

Interpreting dividing by zero becomes even more confusing if we approach it from a different direction. Suppose I want to calculate $4 \div 1/2$. One way to make sense of this is to ask, for example, how many half pizzas can be served up from four whole pizzas? Eight. And $4 \div 1/4$? Sixteen, as there are 16 quarter pizzas in 4 whole ones. So $4 \div 1/8 = 32$, $4 \div 1/16 = 64$, and so on. As we divide 4 by smaller and smaller fractions, we get bigger and bigger answers. As these fractions edge towards zero, our answers gallop towards infinity. Surely it makes sense to say the answer to $4 \div 0$ is infinity?

Which argument is the correct one, the true one? For mathematicians that is not the best question to ask. All these arguments are reasonably logical, depending on your initial premises. The mathematical way out of the conundrum is to simply ban all division by zero. In mathematical terms, division by zero is undefined. This may seem something of a cop-out. Surely there has to be a definitive interpretation for dividing by zero? No, there is not.

So zero is unique – it does not share the status that the other natural numbers have. In some circumstances, zero is treated as a number like any other, but in other circumstances we have to exclude it from that set. No wonder it took many years for mainstream mathematics to accept the inclusion of zero into the mathematical toolkit of numbers.

# ▶ The integers

Many calculations can be done with the natural numbers but many cannot. The answer to $7 - 3$ can be found, but there is no natural number answer to

3 – 7. Now we accept that $3 - 7 = -4$, but despite their great mathematical achievements the ancient Greek, Egyptian and Babylonian mathematicians did not work with negative numbers. Mathematics was a practical science, quantifying and measuring the real world, so the idea of something less than zero was literally unthinkable. The history of western mathematics would have us believe that negative numbers really only came into being around the tenth century, but there is evidence that the Chinese worked with a version of negative numbers several centuries earlier. And in seventh-century India the culture of money-lending led to an arithmetic of 'fortunes' and 'debts': the slow uptake of negative numbers in the west is likely to be due to this association with usury. Even having accepted negative numbers as a useful extension to the natural numbers, mathematicians remained sceptical. For a time the negative numbers were known as *numero absurdo* – absurd numbers – and the finally agreed name – negative – is hardly flattering.

Part of the resistance to negative numbers was probably due to the difficulty in finding some embodiment of them – how do you represent something that seems not to exist? While we meet negative numbers on thermometer scales or in bank balances, it was not until the 17th century that the English mathematician John Wallis came up with the idea of extending the number line back beyond zero. On a number line, addition can be modelled as moving along the line to the right, and subtraction as moving in the opposite direction. So $8 + 3$ and $6 - 3$ look like Figure 2.1.

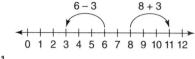

▲ Figure 2.1

Wallis's insight was the power of representing numbers by a position, rather than as a quantity. Extending the number line back beyond zero means a solution to 3 – 7 can now be found that makes sense (i.e fits with our experience of movement forward and back).

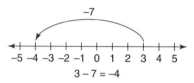

▲ Figure 2.2

# Two negatives make a positive

Before we decide to treat the negative numbers as legitimate we need to check how they behave under multiplication and division, as we do not want to end up with the same sort of contradictions that zero presents.

We know that multiplication is **commutative**, that $a \times b = b \times a$ (see Chapter 4), so if we can find a sensible interpretation of either, for example $3 \times -4$ or $-4 \times 3$, then we can use the commutative law to argue for the reverse to be true. Earlier I 'read' $4 \times 3$ as four multiplied by three, or, four taken three times. That is multiplication treated as repeated addition: $4 \times 3 = 4 + 4 + 4 = 12$.

Reading $-4 \times 3$ similarly, it becomes 'negative four multiplied by three', as $-4 + -4 + -4$. What is a sensible interpretation of adding three negative numbers together? Well, if I imagine borrowing £4 from each of three friends, that is three separate debts of £4, and an overall debt of £12. The sensible interpretation of $-4 \times 3$ is $-4 \times 3 = -4 + -4 + -4 = -12$. By extension of the commutative law, $3 \times -4 = -12$ as well.

Now, $4 \times 3 = 12$ is an example of the pattern that a positive (number) multiplied by a positive is a positive and $-4 \times 3$ is an illustration of the general pattern that a negative multiplied by a positive is a negative. What about that bane of school learners, multiplying a negative by a negative?

Take this sequence of calculations:

$$-4 \times 3 = -12$$
$$-4 \times 2 = -8$$
$$-4 \times 1 = -4$$
$$-4 \times 0 = 0$$

Continuing on, we can extend the calculations and the pattern in the answers. The number that $-4$ is being multiplied by is going down by one each time, so the next two calculations in the sequence are $-4 \times -1$ and $-4 \times -2$. The pattern in the answers is that these go up by 4 each time, so in order to keep everything consistent, the next two equations are:

$$-4 \times -1 = 4$$
$$-4 \times -2 = 8$$

The same logic holds with any pair of numbers and so the sensible solution to the question of what is a negative times a negative is that it is a positive. Teachers often go to great lengths to show why this is so, by making jumps along the number line in reverse, or putting calculations into the context of lifts going up and down, but the fact of the matter is that a negative times a negative is a positive because it keeps the mathematics consistent. It is a generalization, not a representation of anything in the physical world.

Our number system can now sensibly stretch (metaphorically) in two directions ... −4, −3, −2, −1, 0, 1, 2, 3, 4 ... and is renamed: the integers, all the whole numbers, positive and negative, plus zero.

## ▶ Common sense

The mathematics of the natural numbers builds on our common experiences of putting together and taking apart collections of objects and movement forward and back along a line. Once numbers moved away from being adjectives and became mathematical objects studied and explored in their own right, mathematicians were able to look at what was permissible within this system and push the boundaries back. That led to the inclusion of zero as an honorary number, and the development of the integers, a number system including numbers that do not have origins in the real world. The power of mathematics is revealed by the fact that having extended, through invention, the natural number system, it turns out that these made up numbers have powerful applications in the world.

# Visualizing

*I see it, but I don't believe it.*

Georg Cantor

While some mathematicians make sense of the discipline through internal dialogue, most mathematicians report 'seeing things' in their mind's eye, 'visualizing' mathematical ideas. Mathematicians do not see symbols as static images on the page; they see them as coming to life – symbols move, dance and entrance. Symbols are dynamic. A difficulty with mathematics in the written form is that it can lack this sense of the dynamic, so I invite you when reading the following to bring the static images to life in your mind, to create some sense of the play of ideas.

## ▶ Square dance

Imagine two squares, *a* and *b*, of different but fixed sizes. The squares are joined at one of each of their corners, the joint acting as a hinge: the squares can rotate about that point, moving closer together or further apart. Imagine that movement as a continuous motion, from the squares sitting close together with a small angle between them, rotating out to create an intervening wide space, then moving back together again (Figure 3.1).

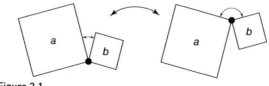

▲ Figure 3.1

Imagine now that these two squares are joined by a third square, *c*. Unlike *a* and *b*, square *c* is a rubbery, stretchy square. It can grow or shrink while always remaining square. This third square attaches itself to the corners of the original squares, bridging the gap between them (Figure 3.2).

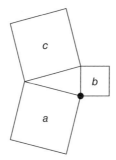

▲ Figure 3.2

As before, squares *a* and *b* rotate together or further apart, and as they do so *c* grows or shrinks accordingly. Again, I encourage you to bring this description to life by imagining *a* and *b* rotating towards and away from each other, with *c* shrinking and growing accordingly (Figure 3.3).

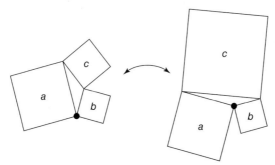

▲ Figure 3.3

In broad terms, as *a* and *b* move and *c* shrinks or grows, what can we say about the relationship between the areas of *a* and *b* together, and how this combined total area compares to the constantly changing area of *c*?

I hope you can see that when the angle between *a* and *b* is small, the area of square *c* is smaller than the areas of *a* and *b* together. At the other extreme, when the angle between *a* and *b* is large, then *c* grows to a size where its area is greater than the combined areas of *a* and *b*: you fit both *a* and *b* inside *c* quite comfortably and still have room over.

The dynamics of this image means that square *c*, with an area less than that of *a* and *b* combined, grows and its area changes until it is greater than that of *a* and *b* together. In this movement between *c* going from having an area that is smaller than the other two squares combined to *c* having an area that is greater than the other two combined (and vice versa), there has to be one moment in that movement where the areas balance. At one point the area of *c* must precisely equal the areas of *a* and *b* combined. It is impossible for the area of *c* to go from being smaller to being greater without passing through this one, and only one, moment of balance, of perfection.

I often present this thought experiment and people's reaction is that it is interesting but most claim never to have seen it before. Pointing out that the moment of balance occurs when the angle between squares *a* and *b* is a right angle makes it seem more familiar to them. If they shift their attention to the space that the three squares surround, they can see that it is, at that moment, a right-angled triangle. The side of the triangle that $c$ is sitting on is the **hypotenuse**, so:

> *The square on the hypotenuse is equal to the sum of the squares on the other two sides.*

There we have it – Pythagoras' theorem (Figure 3.4).

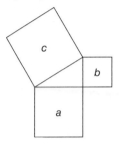

I find this appealing for two reasons. First, it embodies what is for most people an abstract piece of mathematics, as Pythagoras' theorem is most often presented as an algebraic formula:

$$a^2 + b^2 = c^2$$

Not only does this formula fail to conjure up any images, it focuses on the lengths of the sides of the triangle more than it does the squares. In fact, it can be shown that the squares are only one particular case of balance. Everything works just as well if we construct semi-circles on each side of a right-angled triangle – the area of the semi-circle on the hypotenuse will be equal to the sum of the areas of the semi-circles on the other two sides. Or the hexagons, triangles or cat's faces that we could construct. Squares are the canonical choice because their areas are easily calculated.

The second reason this visualization is appealing is that it captures a single moment of calm. There is little we can say about the area of $c$ when the angle between the other two squares is not a right angle. Whoever first found this point of balance managed to pin it down and preserve it, like a butterfly on a board. Rather than Pythagoras' theorem being some dusty formula that needs to be committed to memory and then practised, it takes on a whole new meaning when seen as an instance of tranquillity in a sea of uncertainty.

# ▶ Proving Pythagoras

The astute reader may have noticed that I have glossed over something rather important. Running this mental animation of squares moving establishes that there *is* a moment of balance but it does not establish *when* that moment of balance occurs. It looks as if it happens when the enclosed triangle is right-angled, but can we be certain of this? Another thought experiment, supported by some images, provides a visual proof that this is the case.

Let me abstract out from the images above the triangle and the square $c$ and extend the other two sides a little. If the angle $x$ is a right angle, then we can continue these lines, turn them through right angles at appropriate points and enclose $c$ in a square (Figure 3.5).

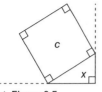

It is not difficult to show that the three new triangles that this construction brings into being are identical copies of the original triangle: it uses the angle-side-angle (ASA) test for two triangles being identical, for being **congruent**. If two triangles each have a side that is the same length, and the angles at the ends of that side are the same, then the two triangles are congruent. The four triangles here have sides the same length – their hypotenuses – as each of these is the length of the side of square *c*. The property that angles meeting on a straight line add up to two right angles (see paragraph after Figure 3.9), together with the fact that the angles of a triangle add up to two right angles, establishes that the angles in each triangle all match. All four triangles are congruent right-angled triangles. So square *c* is surrounded by four copies of the right-angled triangle (Figure 3.6).

▲ Figure 3.6

Now focus on this image slightly differently. Take the larger square constructed around *c* as a fixed frame and suppose that the four triangles are cut from card, so they can be moved around within that frame (Figure 3.7).

▲ Figure 3.7

We can reposition the four triangles so that the 'white space' now shows the squares *a* and *b*. As no white space has been added or removed in this movement of the triangle, the sum of the areas of *a* and *b* must be the same as the area of *c* (Figure 3.8). QED – *quad erat demonstrandum*, or 'which was to be demonstrated'.

▲ Figure 3.8

# ▶ Euclidean geometry

Triangles are the perfect geometrical objects for the mathematician to explore, and typical of classic Euclidean geometry. Simple enough to be studied systematically, rich enough to provide a surprising number of mathematical insights, they are also important practically as constructions based on triangles are rigid. Nail three planks of wood together at their ends to make a triangle and you have a strong, non-deformable shape. Four planks of wood nailed together make a quadrilateral that can be pushed out of 'true'; nailing a plank across the diagonal stabilizes the quadrilateral by creating two triangles. Our built environment looks as though it is made up of squares and rectangles, but it is kept up by triangles. This is most nakedly evident in pylons with their intricate patterns of triangles, so fascinating to some that there is a 'pylon of the month' website. Computer graphics also rely on triangles; behind the magic of characters like Shrek or Woody is a hidden world of computer-generated triangles.

# Euclid

Born around 360 BCE, Euclid is accepted as 'the father of geometry' through his significant work *The Elements*, 13 books about mathematics. Nine volumes are about plane and solid geometry, the other four about number theory. Part of the enduring charm of Euclid's *Elements* is that everything is built on logical argument and this approach – the axiomatic method – is still central to much of mathematics. Starting from a set of only five basic assumptions – postulates – logical deductions are made that build on previous ones and become ever more complex. Solid and steady though this approach is, it does mean that later ideas have to take for granted previously established ones. Euclid does not get round to proving Pythagoras' theorem until proposition 47 of the first book.

Let us look at the example of that proof of the proposition that the internal angles of any triangle add up to two right angles. We can test this by using a protractor to measure the angles (although that is prone to error) or by cutting a triangle out of paper, tearing the corners off and fitting them together (Figure 3.9).

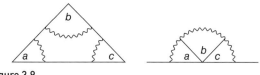

▲ Figure 3.9

But neither of these really proves the claim; they only test it for particular examples. A proof needs to show that the proposition holds true for all possible cases so nothing in Euclid's arguments relies on measurement, by which I mean there are no 360 degrees or metres and centimetres. (Euclid does, however, make extensive use of right angles, and one of his five basic postulates is that 'all right angles are equal to one another'.)

The Euclidean argument builds on two other propositions, which are proved in *The Elements* but which we will have to take for granted, namely that:

▮ angles that meet to form a straight line must total two right angles (180 degrees) – see Figure 3.10

▮ where a straight-line segment crosses a pair of parallel lines, the 'alternate interior angles' are equal – see Figure 3.11

▲ Figure 3.10: Angles meeting on a straight line total two right angles (180 degrees).

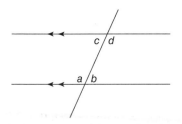

▲ Figure 3.11: Angles *a* and *d* are 'alternate interior angles' and equal, as are angles *c* and *b*.

This is all we need to look at the angles of any triangle. Draw a triangle with its base horizontal (not strictly necessary but it makes it easier to see what is going on), as in Figure 3.12.

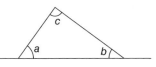

▲ Figure 3.12

Now draw a line parallel to the base of the triangle that just touches the top vertex (Figure 3.13).

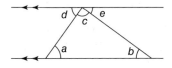

▲ Figure 3.13

Angles *a* and *d* are alternate interior angles, so *a* = *d* and similarly *b* = *e*.

At the top of the diagram we have *c* + *d* + *e*, which must add up to two right angles as these are three angles meeting on a line. Swapping *d* and *e* for *a* and *b* respectively (as they are equal), then *c* + *a* + *b* must also add up to two right angles.

# ▶ Non-Euclidean geometry

Until the 19th century, Euclid's system of plane geometry as set out in *The Elements* was considered the only geometry. When mathematicians started work on other kinds of geometry, based on curved surfaces, the terms 'Euclidian geometry' and 'non-Euclidian geometry' were adopted to distinguish the two kinds of geometry.

The first break from traditional Euclidean geometry came when mathematicians started to look at the properties of shapes on curved surfaces. A boat can travel along three sides of a triangle and end up back where it started; in doing so, it turns through more than two right angles, challenging the Euclidean claim that all triangles have angles adding up to two right angles.

Another extension to geometry came from mathematicians worrying away at Euclid's fifth postulate.

Euclid's first four postulates are:

▶ You can draw a straight line between any two points.

▶ You can extend the line indefinitely.

- You can draw a circle using any line segment as the radius and one end point as the centre.

- All right angles are equal.

All four claims seemed eminently reasonable, and still do. It was the fifth postulate that mathematicians were never completely comfortable with:

- Given a line and a point, you can draw only one line through the point that is parallel to the first line.

Not only is this postulate more cumbersome than the other four, the question was whether it was really as self-evident as the others.

It is beyond the scope here to show the arguments involved, but mathematicians did start to play with the idea that there could be more than one line parallel to another line and **hyperbolic** space came into being.

# Hyperbolic geometry

To imagine hyperbolic geometry in comparison to Euclidean geometry, imagine two straight lines in the two-dimensional plane that are both perpendicular to a third line. In Euclidean geometry these two lines remain at a constant distance from each other, and are known as parallels. In hyperbolic geometry they curve away from each other; these lines are often called ultraparallels.

While mathematicians thought for many years that there was no physical-world model for hyperbolic space, the mathematician Daina Taimina has shown that a piece of crochet that follows a simple rule of a regularly increasing number of stitches does provide one such model. And it turns out that the natural world has an abundance of examples of hyperbolic spaces. Coral reefs are the most striking of these, growing in line with the rules of hyperbolic space long before mathematicians recognized them.

# Generalizing

*The imaginary number is a fine and wonderful resource of the human spirit, almost an amphibian between being and not being.*

Gottfried Wilhelm Leibniz

The ancient game of Go involves placing black or white stones on the intersections of a grid, the aim being to capture your opponent's stones and take command of the board. As with the pieces in draughts (checkers), mancala and a host of other board games from around the world, it makes no sense to ask 'what does that black stone represent?' Go stones do not 'stand for' anything. The stones do serve a purpose in the playing of the game, within the system of rules that has been set up. It does make sense to ask 'what is that black stone doing?'

Our early introduction to mathematics is grounded in situations where mathematical symbols 'stand for' something: the symbol 5 'stands for' a collection of fingers on one hand, a set of nesting Russian dolls, the litter of kittens in a basket. Similarly, the symbol + stands for, we are told, putting together two collections of objects. This representational view of mathematics sticks in our minds so that when, later on, we meet fractions or negative numbers we ask what does 1/2 or −1 'stand for'. But these mathematical objects are like the counters in a game; they do not stand for anything. They are devices, mathematicians' inventions, made to fit with the rules of what already works, to extend existing systems, to stretch rules. They are generalizations that can be traced back to representations of objects or events but ultimately come to have meaning only within the system of mathematics. Thinking of numbers and operations as a system, a game with specific rules, rather than a shorthand representation of things in the world, can help to demystify mathematics.

This chapter looks further at the development of the number system. The word 'development' can imply a natural unfolding – the acorn develops into an oak tree and will never develop into a sycamore – but here I mean development as the way in which mathematicians generalize and create new ideas and symbols to solve problems that at some point in the history of mathematics seemed not to be solvable. This is not a strictly historical account of the development of the number system, but if I pique your interest in that, there are suggestions of where to find out more in the 100 Ideas section.

# ▶ Trouble with division

The set of natural numbers is **closed** for the operations of addition and multiplication, by which I mean that adding or multiplying any pair of natural numbers results in an answer that is another natural number. The set of natural numbers is not, however, closed for subtraction – no natural number provides the answer to 7–12. As we saw earlier, mathematicians came to accept that it was sensible to extend the natural numbers to include negative numbers and zero; the integers came into being.

The extension of the natural numbers into the integers opened up a host of subtraction problems that were previously not solvable, but many divisions of integers remained impossible. We can answer, using our integers, division calculations like $12 \div 3$ or $36 \div 9$, but the integers do not provide answers to $3 \div 12$ or $9 \div 36$. The integers are not closed under division. Mathematicians particularly like closure, so they invented fractions.

# ▶ Fractions

The representational view of mathematical symbols is evident when fractions are presented as objects – this is half an apple, this is a quarter of a pie. One children's picture book I have allows you to flip over split pages to combine fractions and objects: I am not sure what the young learner is meant to make of combinations like a quarter of a bicycle, or a third of a frog.

Mathematically, things like 3/4 of an apple arise from measuring, when the object being measured is smaller than the measuring unit. For example, if the stick ('Unit') illustrated in Figure 4.1 is used to measure the line AB (strictly speaking, a 'line segment' as, for mathematicians, lines extend indefinitely), then without using fractions the best we can say is that the line is a bit shorter than the unit. To be more accurate, the unit can be divided up into fractional parts so that the line can be described as 3/4 of the length of the unit, just as 3/4 of an apple describes the relationship between that quantity of an apple and a whole apple.

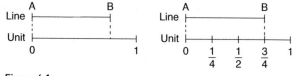

▲ Figure 4.1

Early Greek mathematicians did measure in such a way, but approached it by scaling up rather than subdividing. If a unit or several copies of the unit did not exactly match up with the length of a line being measured, they would take repeated versions of both the unit and the line until the two did line up. Here, three copies of the measuring unit lines up with four copies of the line AB. The unit and line are in the ratio of three to four, or 3/4 as before (Figure 4.2). Hence the mathematical term for fractions as rational numbers – numbers arising out of ratios.

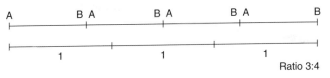

▲ Figure 4.2

All the mathematical activity of the famous Pythagoreans, a community of Greek mathematicians in the sixth century BCE, was carried out in this way using the natural numbers and ratios of them. In general terms, for any measurement activity, the Pythagoreans were convinced that a given unit and a line segment of any length could always be put into a precise ratio of $m:n$. As we shall see when we look at irrational numbers later in this chapter, this turned out to be an incorrect assumption.

## Egyptian fractions

We deal happily with fractions like 3/5 or 7/8 but, like many mathematical ideas, this became mainstream only relatively recently. The Egyptians needed fractions to calculate the construction of the pyramids but, with a couple of exceptions, they worked only with unitary fractions, i.e. fractions with one as their numerator: 1/2, 1/3, 1/5 and so forth. In this system, a fraction like 7/8 can be represented as 1/8 + 1/8 + ... + 1/8, although the evidence is that Egyptian mathematicians looked for ways to record such fractions as concisely as possible, so 7/8 would be 1/2 + 1/4 + 1/8. An engaging mathematical inquiry is to take a fraction, for example 21/23, and to explore the shortest string of unitary fractions needed to express it.

# The number line

Rules, thermometers and speedometers: all embody the idea of a number line, of numbers as positions along a trajectory or path, as a distance travelled from zero. The number line image is a relatively recent mathematical tool, only coming into mainstream mathematics in the 1700s to support the use of negative numbers. Whereas putting the integers on a number line is like placing milestones along a path, extending the image to imagine placing numbers continuously along the path helps us visualize and think about rational numbers. One of the big differences between the integers and rational numbers is the idea of equivalent representations. So the integer two is, mostly, recorded as 2, while a fraction like 2/3 can be recorded as 4/6, 20/30 or an infinite number of equivalent fractions. Before we can confidently put fractions on the number line, we have to be assured that all the equivalences of any fraction all occupy the same place on the line, irrespective of how they were arrived at. That fractions and their equivalences do occupy a unique place on the number line is illustrated by two simple problems.

Suppose I have a bar of chocolate and, in the interests of my waistline, decide to break it into three equal pieces. Lacking willpower, I eat two of the pieces. That I ate 2/3 of the bar can be represented on the number line as shown in Figure 4.3.

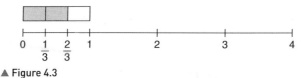

▲ Figure 4.3

Suppose on another occasion, five friends and I share, equally, four bars of chocolate. One way of sharing these is to divide each bar into six equal pieces and for each person to get one piece from every bar. I would then get four one-sixths, and again I can mark this on the number line, rearranging the pieces to show that they are equivalent to 4/6 of a single chocolate bar (Figure 4.4).

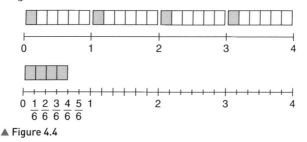

▲ Figure 4.4

It may seem obvious that in each instance I get the same amount of chocolate to eat, but the mathematical equations that model the two situations are quite different. The first is one divided by three and then multiplied by two: $1 \div 3 \times 2$. The second is four divided by six: $4 \div 6$. Our everyday experience tells us that the answer to each of these is going to be the same – putting them on the number line demonstrates that they are also mathematically equivalent. The fraction 2/3 sits in exactly the same spot on the number line as 4/6 (or 20/30 or 12/18 or any fraction equivalent to 2/3). Mathematically, this means that we can be confident that the number line provides a good model not just for the integers but also for the rational numbers; given that any fraction can be expressed in an infinite number of equivalent ways, we are reassured that any equivalent expression will always end up at the same spot on the line.

## Why does the rule for dividing fractions work?

The 'invert and multiply' rule for fractions, as I pointed out in Chapter 1, can seem mysterious or even nonsensical. But meaning can be brought into being by looking at how we read the division symbol. Take a calculation like $35 \div 5$. What sort of simple problem comes to mind? Often a story about sharing: 'I have 35 apples to share equally between five friends. How many apples does each friend get?' Such stories are grounded in our childhood experiences of sharing: one for you, one for you, one for me, and so on. Sharing becomes our default interpretation of division, but that makes it difficult to make sense of dividing by a fraction. How does $6 \div 2/3$ translate into a sharing problem? It does not make a lot of sense to talk about six apples shared between 2/3 people.

There is a second, less often remembered, interpretation of division. I have 35 apples that I want to put into bags of five. How many bags can I fill? That can also be calculated through 35 ÷ 5 and the answer is again 7. In each case we have to ask 7 what? In the first example the answer is 7 apples – the number of apples each of the five friends gets. In the second example the answer is 7 bags (each containing 5 apples). This second example is division as repeated subtraction, and it is this interpretation of division that helps us make sense of division by fractions: how many times can I subtract 2/3 from 6 is 6 ÷ 2/3. To answer that we need to find out how many thirds there are in 6: since there are three-thirds in one, there will be 6 × 3 thirds in 6. To find out how sets of two-thirds can be made from these 18 1/3s, we divide by 2. Combining these two steps, 6 ÷ 2/3 = 6 × 3 ÷ 2, which we can write as 6 × 3/2. We took one calculation that looked difficult to compute, 6 ÷ 2/3, and replaced it with an equivalent calculation that we can compute, 6 × 3/2. This underlying equivalence is hidden within the invert-and-multiply rule, a rule that seems to be about manipulating digits, a rule that you have to take as given and remember rather than understand. Mathematics thus becomes meaningless, when it need not be.

Which brings me back to 1/2 ÷ 1/3. Think of this as 'how many times can I subtract a third from a half?' and we can see that one whole third can be taken away from a half, leaving 1/6, which can be thought of as half of a third, so I can subtract 1/3 from 1/2 one and a half times.

The answer to 1/2 ÷ 1/3 in this reading is 1½ thirds – the original half has not miraculously grown to become one and a half, it has simply been re-presented as one whole third and one half of a third.

## Zooming in on fractions

Imagining the rational numbers as positions on the number line allows us to play around with the image. Taking any two fractions, we can zoom in on that part of the line. Take, for example, 5/16 and 6/16 (see Figure 4.5).

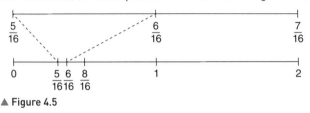

▲ Figure 4.5

Zooming in, literally or imaginatively, we can see that in the space between these two fractions there is plenty of room to squeeze in another fraction. The fraction 5.5/16 fits neatly between the two, but as we tend not to record fractions with decimal points in them (although there is no mathematical law against that) we can record that fraction as 11/32. Converting the original two fractions to 32nds we have 10/32 and 12/32, and 11/32 clearly sits between these. We can now take 10/32 and 11/32 and squeeze another fraction between them; 21/64 sits in the middle. In theory we can continue this forever; given any two fractions, however small the gap between them, we can squeeze another fraction in between. Our number line feels packed with fractions, as indeed it is.

Intuitively it would seem that the number line is so densely packed with fractions that there can be no room for any other numbers to be squeezed in. This intuition is not correct, however, as the Pythagoreans found out.

# ▶ The irrational numbers

Difficulties arose when the group of Greek mathematicians associated with Pythagoras started looking at how to measure the diagonal distance across a square with sides of one unit length. As we saw in Chapter 3, from Pythagoras' theorem this line, being the hypotenuse of a right-angled triangle of side 1, has length $\sqrt{2}$ (see Figure 4.6):

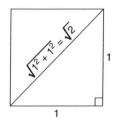

▲ Figure 4.6

The Greek mathematicians were convinced that, using the logic of ratios to measure as described above, there must be some unit (1) a multiple of which would exactly fit along a line that was a whole number multiple of $\sqrt{2}$. That is, whole numbers $n$ and $m$ could be found that would make this equation true:

$n \times \sqrt{2} = m$

The fact that no one could come up with values for *n* and *m* was simply an indication, they thought, that the search had to continue. Legend has it that the Pythagorean mathematician Hippasus finally came up with the definitive argument to show why whole number values of *n* and *m* would never be found, using a classic mathematician's argument, proof by contradiction or *reductio ad absurdum*.

The starting point is to make the assumption that what you want to show is true – so suppose there are natural numbers *n* and *m*, so that:

$$n \times \sqrt{2} = m$$

We can rearrange this a bit to:

$$\sqrt{2} = \frac{m}{n}$$

Thus, our starting assumption is that $\sqrt{2}$ can be written as a fraction. We can also take it that if *n* and *m* have any common factors, we can divide the numerator and denominator by these and thus start with a fraction where *m* and *n* have no common factors (so if the fraction was, for example, 15/35, we express this in the equivalent form of 3/7 by dividing the numerator and denominator by their common factor of 5).

If $\sqrt{2} = \frac{m}{n}$

Then $2 = \left(\frac{m}{n}\right)^2$ [squaring both sides of the equation preserves the equality]

$$2 = \frac{m^2}{n^2}$$

So $2n^2 = m^2$ (multiplying both sides of the equation by $n^2$).

If $2n^2 = m^2$, then both of these numbers are even. If $m^2$ is an even number, then $m$ must be even (squaring an odd number would result in an odd number). If $m$ is even, we can replace it with $2p$.

$$2n^2 = (2p)^2 = 4p^2$$

Dividing both sides of the equation by two gives us:

$$n^2 = 2p^2$$

By the same logic that showed that $m$ has to be an even number, this tells us that $n$ must be an even number. So if $\sqrt{2}$ can be expressed as our original fraction $m/n$, we have shown that $m$ must be even, and $n$ must also be even. In other words, $m$ and $n$ share a common factor of two. But we started off by assuming that $m$ and $n$ had been chosen so that they had **no** common factor. The steps in the argument are all watertight but we have reached a logically impossible conclusion, an absurdity. The only logical conclusion we can make is that the initial premise cannot be true. The square root of 2 cannot be expressed as a fraction.

The Pythagoreans were so distressed by this challenge to what they believed to be true that, so the story goes, they had Hippasus drowned at sea.

# Why irrational?

What was it about $\sqrt{2}$ not being expressible as a fraction that so disturbed the Pythagoreans? Our number line image provides some sense of the challenge that this insight presented. Recall that, intuitively, we can think of the number line as packed with rational numbers. It would seem so densely packed with the rational numbers that there can be no space left on the line into which any other numbers can be fitted.

The startling, counter-intuitive fact that the rational numbers do not fill the number line completely is demonstrated by a simple construction. Setting up a number line and constructing, as before, a right-angled triangle with the perpendicular sides of unit length creates a hypotenuse with length $\sqrt{2}$. Setting compasses to that length, this distance can be marked on the line (see Figure 4.7). So $\sqrt{2}$ certainly exists as a point on the line (in treating this as a thought experiment, considerations like the thickness of the pencil mark are set aside).

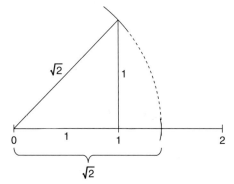

▲ Figure 4.7

A number that cannot be expressed as a fraction has thus been squeezed onto a line that already felt as though it were solidly filled with fractions, a line that had no space for any other numbers. And $\sqrt{2}$ is not unusual – there are infinitely many of these non-rational numbers that can be positioned on the line. As I explore in Chapter 9, one of the joys (or challenges, depending on your disposition) of mathematics is that mysterious, puzzling things happen at the extremes – the infinitely large or, as in this case, the infinitely small – that challenge our common sense expectations, and it was this that the Pythagoreans found so difficult to accept. Mathematicians eventually named these numbers the irrationals, which, while drawing attention to the fact that they cannot be expressed as rational numbers, does still convey the idea of them being somewhat mad.

To recap, our number system grew from the counting numbers to the integers (positive and negative whole numbers and zero), then to the rational numbers, and then to the rational numbers plus the irrational numbers – this full set becoming known as the **real numbers**. And that, so the mathematical community thought, must mark the complete set of numbers. Until, that is, **imaginary numbers** came along.

# Square root of minus one

The real numbers allow mathematicians to solve almost any calculation, but one that the mathematical imagination took time to embrace solving was $\sqrt{-1}$. Multiplying any two real numbers together had always to result in a positive number, so there could never be a number multiplied by itself that resulted in −1.

Yet mathematicians had come to accept that negative numbers – with no real-world counterparts – could come to be treated as proper numbers, so why not act as though $\sqrt{-1}$ was a number, one that could solve previously unsolvable problems? The resistance to $\sqrt{-1}$ rests in two issues. First, if $\sqrt{-1}$ was accepted as a number, would that open the door to a continuing stream of new numbers being invented? Second, given that $\sqrt{-1}$ could not be represented on the number line, how could it be imagined? Gauss, in his doctoral thesis of 1799, took up both of these challenges. First, Gauss proved that if $\sqrt{-1}$, or *i* as it is now known, was accepted as a number, then *all* possible equations could then be solved – there would be no need to invent any more numbers.

With the acceptance of *i* as a number, then any equation can be solved, using a number expressed as a sum of a real number and a multiple of *i*, for example $3 + 5i$, subsequently known as **complex numbers**. Gauss' second insight was to extend the number line into two dimensions and to make the metaphorical link between complex numbers and the system of map coordinates. The complex number $3 + 5i$ could thus be represented by a point 3 units along the *x*-axis and 5 units up the *y*-axis. Although this image – now called the Argand diagram – is universally used to work with complex numbers, Gauss actually hid his use of it, removing it from the final argument. So strong was the feeling at the time amongst mathematicians that images could deceive, that Gauss did not admit to using these images until forty years later.

# Pattern sniffing

*Out of an infinity of designs a mathematician chooses one pattern for beauty's sake and pulls it down to earth.*

Marston Morse

Pattern sniffing is a core mathematical habit of mind: mathematicians are always on the lookout for a pattern or two. For example, what comes to mind with the number 64? 'Not a lot' may be an immediate response, but then you might recall that there are 64 squares on an eight-by-eight chessboard. Sixty-four is thus a square number – that number of pebbles can be laid out to form a square array (see Figure 5.1).

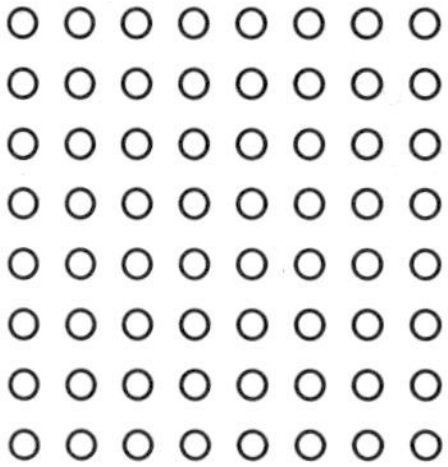

▲ **Figure 5.1**

The mathematically curious look at this arrangement and might see contained within it the other square numbers from 1 × 1, 2 × 2 up to 7 × 7. Systematically marking these off reveals another pattern – that the square numbers successively increase by the odd numbers (Figure 5.2):

$1 = 1 \times 1 = 1$

$4 = 2 \times 2 = 1 + 3$

$9 = 3 \times 3 = 1 + 3 + 5$

$64 = 8 \times 8 = 1 + 3 + 5 + 7 + 9 + 11 + 13 + 15.$

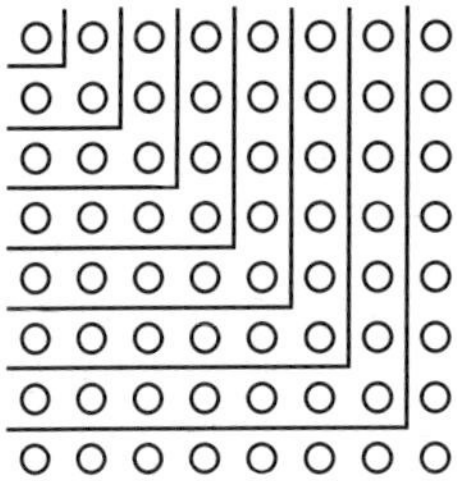

▲ Figure 5.2

Slicing the square along a diagonal and without splitting the pebbles exposes another pattern – that 64 is made up of two sums of consecutive numbers, 1 + 2 + 3 ... + 8 and 1 + 2 + 3 ... + 7 (see Figure 5.3). These sums of consecutive numbers are the **triangular numbers**, so named because our pebbles can be laid out in a triangular representation. Sixty-four is the sum of two consecutive triangular numbers, a fact that holds true for any square number. For example, the fourth triangular number is 1 + 2 + 3 + 4 = 10 and the fifth is 1 + 2 + 3 + 4 + 5 = 15, and 10 + 15 = 25 = 5 × 5.

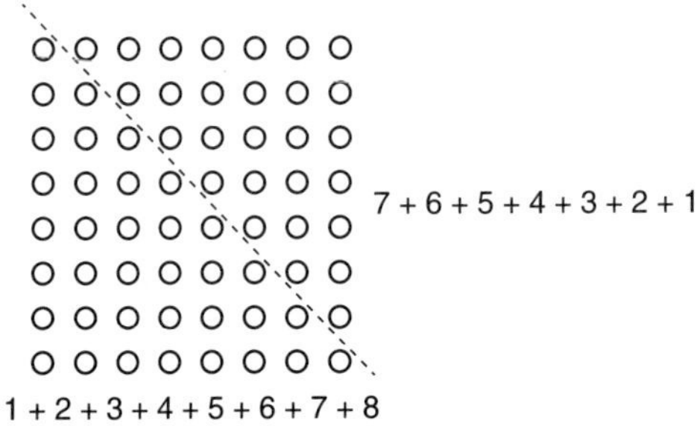

7 + 6 + 5 + 4 + 3 + 2 + 1

1 + 2 + 3 + 4 + 5 + 6 + 7 + 8

▲ Figure 5.3

Sixty-four cries out, to the mathematician, to be repeatedly halved – 32, 16, 8, 4, 2, 1 – which in reverse shows us that 64 is two doubled five times: 2, 4, 8, 16, 32, 64, or $2 \times 2 \times 2 \times 2 \times 2 \times 2$ or $2^6$. This expansion of 64, with some brackets put in, brings us back to it being a perfect square: $(2 \times 2 \times 2) \times (2 \times 2 \times 2) = 8 \times 8$. Putting brackets elsewhere we can have $(2 \times 2) \times (2 \times 2) \times (2 \times 2)$ $= 4 \times 4 \times 4$, revealing that 64 is a perfect cube – sixty-four sugar cubes can be stacked to make a $4 \times 4 \times 4$ cube.

Sitting between 63 and 65, 64 begins to look considerably more interesting than its neighbours. No wonder the Beatles sang about it.

# ▶ The rules of arithmetic

All these 'facts' about 64 are expressed through our system of arithmetic and the basic operations of addition and multiplication. I say 'basic', but mathematicians are fascinated by the underlying structures, similarities and differences in these operations, as illustrated by (re)visiting some laws that you are likely to have met before, even if only intuitively.

Earlier I took it as a given that addition is commutative: for any addition, reversing the order of the numbers does not affect the answer. For example:

$$345 + 199 = 199 + 345.$$

It is not difficult to show why this is true. While we could calculate the sum for each side of the equation, we know from everyday experience that a pile of 199 pebbles added to an existing pile of 345 pebbles is going to give the same result as starting with a pile of 199 and adding another 345. In general:

$a + b = b + a.$

What about $345 \times 199 = 199 \times 345$? This is also true, but we may pause and begin to wonder if we are as certain as for addition. Multiplication is usually taught as repeated addition: 345 multiplied by 199 means adding 345 to itself 199 times. Is it obvious that taking 199 and adding it to itself 345 times is going to give the same result?

The mathematics educator John Mason talks about different types of convincing: convincing yourself, convincing a friend, convincing a sceptic. Convincing yourself means being content that the consistency of everyday experiences means having no reason to doubt that they will continue to be consistent. If every experience we have had of reversing the order of multiplying two numbers has always given the same answer, there is no reason to suppose $345 \times 199$ and $199 \times 345$ should behave any differently. If a friend needed convincing, checking on a calculator would suffice. 'Ah yes,' says the sceptic, 'but *why* is that the case? How can you know for certain that this will hold true in all possible cases?' A different image from multiplication as repeated addition is needed. Recall 64 as a perfect square: established by showing $8 \times 8$ as an array of pebbles, 8 in each row and 8 in each column. Although it would be ridiculously large to set up, we can imagine $345 \times 199$ as an array of 345 rows of 199 pebbles. Rotating this array through a quarter turn, it is then 199 rows of 345 pebbles: 199 × 345. Whichever way we 'read' the array, the total number of pebbles does not change, something that will hold true irrespective of the number of pebbles in each row or column. So if the array was *a* pebbles by *b* pebbles, then:

$$a \times b = b \times a$$

The name, the **commutative law**, has the same origins as 'commuting' as in going back and forth, from the Latin *commutare* 'to often change, or change altogether' – a little ironic as what is being drawn to the attention here is the lack of anything changing if you reverse the order of calculation.

Why do mathematicians call this a law? In general, we use the word 'law' in two different senses. There are laws that describe behaviour and laws that prescribe behaviour. The law of gravity describes how dropped objects will always accelerate at the same rate, whereas the law of the road in the UK prescribes that vehicles should be driven on the left. In the former case, a generalization is observed; in the latter case it is created. The commutative law is a mathematical description of behaviour that could not be otherwise: in the world of mathematics, $a + b$ will always be equal to $b + a$ and $a \times b = b \times a$.

In one of Moliere's plays, a character, Monsieur Jourdain, says he has 'been speaking prose all my life, and didn't even know it!' Similarly, most people get through life using the commutative law in their everyday calculations without thinking about it, but as we will see, it is important to the mathematician that it be explicitly articulated in this way. It is also important to know when a law like this does not apply. In general, while addition and multiplication are commutative, subtraction and division are not ($\neq$ means 'not equal to'):

$17 - 5 \neq 5 - 17$

and

$17 \div 5 \neq 5 \div 17$

## In general

A tricky aspect of mathematics is that mathematicians can use phrases with a subtly different meaning from their everyday use. 'In general' is one such phrase. We would not be called to account for saying things like 'In general, men are taller than women' or 'In general, women live longer than men'. In each instance it is the case that the average height of men is greater than that of women, and the average life expectancy (in some countries) of women is greater than it is for men. But we also accept that some women are taller than some men, and that some men live longer than some women. However, when a mathematician says 'in general', they mean something holds for *all* possible cases.

Take another pair of simple calculations. How do you calculate the answers?

$12 + 3 + 5$

$7 \times 25 \times 4$

For the addition, you will have seen that the answer is 20, but did you arrive at that by adding $12 + 3$ and then adding 5 to 15, or by seeing that $3 + 5$ is 8 and knowing this would make 12 up to 20? The order of adding makes no difference to the final total, and the same holds for multiplication. Noticing that 25 × 4 = 100 and then multiplying that by 7 is easier than calculating 7 × 25 and then multiplying that answer by 4. When adding or multiplying more than two numbers the calculation has to be carried out by operating on one pair of numbers at a time, and the important observation is that it does not matter which pair we choose to start with:

$$(12 + 3) + 5 = 12 + (3 + 5)$$

$$(7 × 25) × 4 = 7 × (25 × 4)$$

Again, we know that these give us the same answer, but although this seems instinctively obvious, that is not the case in most of our life. Making an omelette involves cracking eggs into the bowl, whisking them, and cooking them – crack + whisk + cook – but (crack + whisk) + cook is not the same as crack + (whisk + cook); you cannot whisk and cook the eggs before cracking them. The flexibility of the order of calculating a string of additions or products is a regularity that mathematicians call the **associative law**, expressed in its most general terms as:

$$(a + b) + c = a + (b + c)$$

and

$$(a × b) × c = a × (b × c).$$

# The commutative and associative laws together

Given 36 + 17 + 14, the savvy calculator will spot that it is easier to add 36 and 14 first and then to add on

17 than it is to carry out the calculation in the order set out. Again, this seems obvious, but mathematically the steps taken involve first applying the commutative law, so that 36 + 17 + 14 becomes 36 + 14 + 17 and then applying the associative law to the order of calculating (36 + 14) + 17.

In fact, there are 12 different ways that 36 + 17 + 14 could be calculated:

| | |
|---|---|
| (36 + 17) + 14 | (17 + 36) + 14 |
| 36 + (17 + 14) | 36 + (14 + 17) |
| (17 + 14) + 36 | (14 + 17) + 36 |
| 17 + (14 + 36) | 14 + (17 + 36) |
| (14 + 36) + 17 | (36 + 14) + 17 |
| 14 + (36 + 17) | 17 + (36 + 14) |

And thanks to everyone's innate sense of pattern, we most often choose the most effective method.

Add four numbers and there are 120 different ways to do this. It is thanks to the commutative and associative laws that we can be confident that all 120 answers would be the same.

# The distributive law

There is one other law, the **distributive law**, which links addition and multiplication. For example, calculating 35 × 7 by doing 30 × 7 and then adding on 5 × 7 uses the distributive law, which in its general form is:

$(a + b) \times c = (a \times c) + (b \times c)$

And that is it. The commutative, associative and distributive laws concisely express all the legitimate moves we can make in any calculation involving addition and/or multiplication.

# ▶ Prime numbers

Sixty-four is an example of a **composite number** – 64 pebbles can be set out in an array which is more than one pebble wide. Eight by eight is one way, as is 32 by 2 or 16 by 4. But 67 pebbles can be laid out only in an array that is one pebble wide: 1 × 67. Sixty-seven is a **prime number**.

Prime numbers are the genes of the number system: all other whole numbers can be expressed as a product of primes: 65 = 5 × 13, 66 = 2 × 3 × 11. Many people recall that a number is prime if it can only be divided by 1 and itself, if its only factors are 1 and the number itself (factors are the whole numbers that will divide exactly into a whole number and leave no remainder). The only factors of 67 are 1 and 67.

An alternative definition is that a prime number has exactly two factors. Isn't that what the first definition implies? The factors of 67 are 1 and 67 – exactly two of them. There is, however, one crucial difference: under the first definition, 1 is a prime number – it is divisible by 1 and by itself (which happens also to be 1). But 1 only has one factor – 1 = 1 × 1 – it is not the product of two *different* factors. Under the first definition 1 is prime, under the second it is not. So which definition to go with? The second definition was agreed on by mathematicians precisely to exclude 1 from the club of primes. Why? It all boils down to the Fundamental Theorem of Arithmetic – you can tell mathematicians think this is important by the pomposity of the name.

This 'fundamental' theorem says that any number can be expressed as a *unique* product of primes. Take, for example, 12. To express that as a product of primes, we factorize it into a product of any two numbers, for instance $2 \times 6$. Two is prime, but we can factorize the 6 into another product, $2 \times 3$. So as a product of primes $12 = (2 \times 6) = 2 \times (2 \times 3) = 2 \times 2 \times 3$. The fundamental element is that we always end up with this factorization irrespective of the initial factoring: $12 = 4 \times 3 = (2 \times 2) \times 3 = 2 \times 2 \times 3$, as before. This works for any number. Accepting 1 as a prime number would be a fly in this pure ointment; 12 could then be $1 \times 2 \times 2 \times 3$ or $1 \times 1 \times 1 \times 1 \times 2 \times 2 \times 3$ or an infinite number of products of primes; its prime factorization is no longer unique. Lonely though it may make 1, without this fundamental theorem being made to hold true, advances in branches of applied mathematics such as crystallography and codes would have been severely hampered.

## The power of the primes

Prime numbers are enormously useful. The simplicity of the primes – having no factors less than themselves (except 1) – creates a challenge. Looking at a number's factors is a way into investigating that number (the beginning of this chapter would have been much duller if I had chosen to start with 67 rather than 64), making prime numbers themselves difficult to analyse. But this is also their strength – the difficulty in tracking down prime numbers has made them the bedrock of Internet security. Enormous prime numbers encrypt electronic data, making it extremely difficult for anyone to decode it.

Prime numbers are intriguing to mathematicians. Why? Because the pattern behind them has yet to be found. The only way to find the complete set of prime numbers up to a given ceiling is to list them all. For example, the prime numbers less than 50 are 2, 3, 5 , 7, 11, 13, 17, 19, 23, 29, 31, 37, 41, 43 and 47 – there is no pattern here that helps predict what the next prime number will be. Having found any prime number, there is no predicting when the next one will occur without checking all the numbers as you go on, and mathematicians have yet to find a way to predict how many prime numbers there are up to a given number. In 1859, the mathematician Bernhard Riemann did propose a hypothesis concerning the behaviour of the primes, and it remains one of mathematics' great unsolved puzzles, with a $1 million prize awaiting the mathematician who comes up with a proof.

## ▶ Common sense

The examples of pattern that I have illustrated with 64 are, in a sense, common to us all; my playing and laying out pebbles has to produce the same arrangements that a bored cave-person might have stumbled upon – in fact, the word 'calculate' comes from the Latin for a small stone. Much of mathematics is based on such shared, common-sense making. Except, of course, when it isn't. Let us look at where the mathematics can diverge from common sense and why it does so.

A classic 'I can't make sense of it' piece of mathematics is whether or not 0.9999 … is equal to 1 (the ellipsis indicates that the decimals continue infinitely). Common sense would suggest that it is not; while the more decimal places that are added to 0.9999 …. the closer it must get to 1, it will never actually reach 1. If we think about this through the metaphor of moving along a line, as we creep towards 1, the steps are getting (infinitely) smaller and the end point looks as though it is in reach but it will always be beyond our grasp.

Arguments in favour of 0.9999 …. being equal to 1 shift gear into logical deduction. For example, we accept that $1/3 = 0.3333$ …. And since $3 \times 1/3 = 1$, then $3 \times 0.3333$ …. $= 1$. That is, 0.9999 … must also be 1. Perhaps all this argument does is to hide another contrary-to-common-sense result. The fraction 1/3 is turned into the decimal fraction by dividing 3 into 1. The routine for this goes 'three into one won't go, put down a zero, add a decimal point and turn the remainder of one into ten-tenths. Three into ten-tenths goes three times, remainder one-tenth, turn that into ten-hundredths' and so on. At whatever point we stop the division, there is always a remainder of one to be divided up into ten smaller parts. In other words, common sense tells us that 0.3333 ... is a bit short of being equal to 1/3, so multiplying that by three means 0.9999 ... is short of equalling one. Ah, the mathematician says, but algebra proves the case. Let's use $x$ to stand in for the value of 0.9999.

$x = 0.9999$ ...

Multiply both sides of this equation by 10: $10x = 9.9999$ ... (multiplying a number by 10 moves the decimal point one place to the right – I know some purists prefer to say that the decimal point is fixed and the digits moves, but the end result is the same.)

Expanding this a little:

$10x = 9 + 0.9999$ ...

Or $10x = 9 + x$

Therefore $10x - x = 9$, or $9x = 9$, or $x = 1$. Common sense is trumped, as far as the mathematician is concerned, but for many people, this argument appears as convincing as being told that the hat from which the rabbit was pulled was previously empty.

The conundrum of whether or not 0.9999 .... = 1 was finally resolved in the early 19th century when the mathematician Augustin-Louis Cauchy developed the idea of limits. It is beyond the scope of this book to go into detail about this here, but the basic idea revolves again around the distinction between symbols standing for something and defining what a symbol means. While the symbol 9, in a sense, stands for all those times when we have experienced 'nine-ness' (be it nine cats, tails or lives), the decimal 0.9999 ..., being infinite, cannot stand for the sum of such experiences. We can imagine a decimal with an infinite number of places, but no one has ever experienced the entire number. So we can choose what we want 0.9999 ... to stand for; we can define it to mean what we would like it to be. This does not, of course, mean defining it to be what we like. The breakthrough that Cauchy made was to show that given the choice between defining 0.9999 ... as not equal to 1 and defining it as equal to 1, then the latter choice is the more mathematically *useful* one. In short, 0.9999 ... is defined to be 1 because that works best.

# Mathematicians accept what works

Another example illustrates this further. Take $x^{-1}$. This is defined as being equal to $1/x$. So $2^{-1}$ is $1/2$, $5^{-1}$ is $1/5$ and so on. Since the fractional notation of $1/2$, $1/5$ and so forth would seem sufficient for most purposes, why go to the bother of introducing another notation, one that just appears to complicate matters? The answer comes from the mathematician's mental habit of looking for patterns, looking for generalizations, looking for connections. The starting point is **indices** – the shorthand way of recording a number repeatedly multiplied by itself: $10 \times 10 = 10^2$, $4 \times 4 \times 4 = 4^3$ and so on. In general terms $n^p$ is a string of $p$ $n$s multiplied together, described as $n$ raised to the power of $p$.

Now suppose we want to multiply together, for example, $10^3$ and $10^4$? $10^3 \times 10^4 = (10 \times 10 \times 10) \times (10 \times 10 \times 10 \times 10) = 10 \times 10 \times 10 \times 10 \times 10 \times 10 \times 10 = 10^7$.

A few such examples and the pattern becomes clear. We don't have to write out the full calculation, we can simply add the values of the powers: $7^5 \times 7^6 = 7^{11}$, $2.7^8 \times 2.7^{10} = 2.7^{18}$. Algebraically: $n^p \times n^q = n^{p+q}$. At some point mathematicians started to ask, what if we had $10^{-1}$? What would be a sensible meaning for that? Common sense lets us down; 'multiply ten by itself negative one times' does not seem meaningful. But a rule has been established – to multiply two powers of the same number simply add the indices. A neat rule, worth preserving. In that case, applying the rule to $10^3 \times 10^{-1}$, the answer has to be $10^{3+(-1)} = 10^2$. Multiplying $10 \times 10 \times 10$ by $10^{-1}$ results in $10 \times 10$; in other words, multiplying by $10^{-1}$ has the same effect as dividing by 10. When multiplying, what number has this effect of dividing by 10? $1/10$. $10^{-1}$ is defined as $1/10$, $n^{-1}$ is defined as $1/n$ so as not to damage the rule of adding the powers. It is a meaningful definition (in that it preserves the coherence of mathematics) rather than a sensible one (in appealing to our sense of having origins in the 'real' world).

# Proving

*If only I had the theorems!*
*Then I should find the proofs*
*easily enough.*

*Bernhard Riemann*

Proof lies at the heart of mathematics. When totting up a list of expenses, it may come to your attention that adding two odd numbers regularly results in an even answer. Thinking back, you realize that this has always been the case and it is just obviously 'true'. Noticing such regularity is only the beginning of the story for the mathematician, who would describe the statement that 'adding two odd numbers makes an even number' as a 'conjecture', a **conjecture** being a statement of generality that has yet to be established as always holding true or not. The sceptical mathematician (who in everyday life will happily accept that adding two odd numbers results in an even number) would say, 'You can never have checked them all, so how can you be certain that there isn't a pair of odd numbers somewhere that added together give you an odd number?' One reply to this might be 'Why don't you get a life?', but the desire for completeness, for closure, is a mathematical habit of mind leading to the desire for a proof.

For this particular conjecture, a 'proof' is easy; we can do it on the back of an envelope, without algebra, if we accept that an odd number can be represented as illustrated in Figure 6.1. The 'broken' section indicates that the 'bar' can be made as long as we like, so this image can represent any two odd numbers.

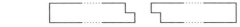

▲ **Figure 6.1**

Putting two of these 'bars' together, we can easily see that the 'protrusions' fit and so the result must be an even number (Figure 6.2).

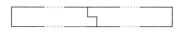

Turning this into algebra, any odd number can be expressed as $2n + 1$ (which is just the image above presented in letters, with $n$ standing for the length of the double bar). Adding two such numbers: $(2n + 1) + (2m + 1) = 2n + 2m + 2 = 2(n + m + 1)$, which is an even number.

This form of direct, deductive reasoning is one type of proof, where a number of premises are taken as given (in this case that $2n + 1$ represents any odd number) and then linked to a conclusion through a series of logical steps. In presenting a proof, certain assumptions have to be made about how much detail to include. The step from $(2n + 1) + (2m + 1)$ to $2n + 2m + 2$ is valid through using the commutative and associative laws described in Chapter 5. Spelling out the steps looks like:

$(2n + 1) + (2m + 1) = 2n + 1 + 2m + 1$

$= 2n + 2m + 1 + 1$ (using the commutative law)

$= (2n + 2m) + 2$ (using the associative law)

I chose not to spell out these steps earlier on the assumption that the reader would be happy to accept the argument in one step. Every mathematical proof makes such assumptions about steps assumed to be self-evident. That is why a proof can seem so opaque to a reader – the prover has to make assumptions about how much the intended reader will take as obvious but that will always vary from reader to reader. There is an apocryphal story of a mathematics professor presenting a proof in a lecture and in moving from one step to the next, saying 'It obviously follows that. Hmm, at least, I think it obviously follows. Excuse me.' and then leaving the room. The professor returns 20 minutes later, announcing, 'Yes, it does obviously follow.'

Another difficulty with proofs is that they hide the messiness that mathematicians go through in constructing them. A myth around proofs, maintained by how we meet them in school, is that mathematicians just think them up on the spot, that a proof pours out of the brain and onto the page complete and coherent. While the preliminary work behind great paintings is made public through access to artists' sketchbooks and the 'roughs' that precede final versions, we rarely, if ever, have access to a mathematician's 'roughs'. Proving mathematics by presenting a series of well-constructed statements gives as much insight into the processes of proving as reciting a poem does into the processes of composing it.

# ▶ Composing a proof

Just as symphonies do not pour out of the composer's mind fully formed, so too a mathematical proof has to be 'composed' – drafted, refined, redrafted, until the steps are arranged as harmoniously as the notes in a symphony. And as symphonies have a different form from concertos, so too there are various classical forms that proofs take. We met one earlier – proof by contradiction – when looking at why $\sqrt{2}$ could not be expressed as a fraction. For some, proof by contradiction feels a little like cheating – basing a positive outcome on a negative conclusion seems, well, contradictory. Proof by deductive reasoning and proof by induction are more constructive in their approaches so let us look at the processes of composing each of these types of proofs through working on a particular problem. You will probably get more out of reading what follows if you spend time working on the problem before reading on, so I encourage you to get out pen and paper and take a few minutes to play with the problem.

# ▶ The handshakes problem

There are 20 people at a party. Everyone shakes hands with everyone else – once only. How many handshakes take place? (Mike shaking hands with Mary counts as one handshake).

Welcome back if you went away to play with this. Bear with me if you didn't.

As posed, this is a closed problem – there is one specific answer – so nothing needs proving as such, just a convincing argument put forward for the specific answer. Proving comes from looking to see if there is a general rule. I could have posed the problem in general terms – if there were $n$ people at the party, find and prove the general rule for calculating the total number of handshakes. Such a presentation of the problem is, however, initially more off-putting; getting a feel for what is going on is best done by working with a specific number of guests, and 20 is as good a number as any to start with.

One way of finding the answer is to work systematically through the number of handshakes made by each guest. We can set up a mathematical model of the guests by numbering them from one to 20. Working from the end of the line, Guest 20 will shake hands with 19 people. Having done that, she can step aside as she has done her share of the handshaking. Guest 19 shakes hands with the remaining 18 guest and steps aside. Guest 18 will make 17 handshakes, and so on. Shaking hands in this organized fashion, we see a pattern emerge: each guest will shake one less number of hands than their position in the queue. So altogether there will be 19 + 18 + 17 .... + 3 + 2 + 1 handshakes. (Just to confirm what is happening at the end, Guest 2 will make one handshake and Guest 1 has no further hands to shake, so I could have added '+ 0' to the end so that the handshakes of all 20 guests were listed). There is now a fairly tedious calculation to carry out but we can get the answer – 190 handshakes.

Core mathematical habits of mind are to look for efficient methods and generalizations, so the mathematician would not be content to leave the problem at that point, as two questions are emerging:

▶ Is there a shortcut method to finding the total?

▶ Can we come up with a way of easily finding the total for any number of partygoers, 50, 200 or *n*?

Taking the first question, can we find an efficient way of getting the total? The key here is not to take the calculation as set up: 1 + 2 + 3 + ... + 18 + 19 but to play with this to make the calculation easier. If we write out the full calculation – 1+2+3+4+5+6+7+8+9+10+11+12+13+14+15+16+17+18+19 – we might notice that the numbers could be added in a pairwise fashion, from the outermost pair working in (Figure 6.3).

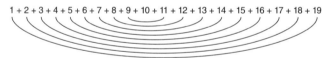

1 + 2 + 3 + 4 + 5 + 6 + 7 + 8 + 9 + 10 + 11 + 12 + 13 + 14 + 15 + 16 + 17 + 18 + 19

▲ Figure 6.3

There are nine pairs that add to 20: 1 + 19, 2 + 18, 3 + 17, ... 9 + 11. That is a total of 180 plus the 10 that is unpaired, making 190. This works, but what happens when there is a even number of numbers to add up, when there will not be that single number in the middle of all those pairs? If there were 21 people at the party, this adding pairwise will give us pairs totalling to 21 and there will be ... it all starts to get a bit confusing.

How else could we play with adding 1 to 19? Looking again at the sum written out in full, we might notice that adding it twice is neat, if we write out the second version in reverse.

1+2+3+4+5+6+7+8+9+10+11+12+13+14+15+16+17+18+19

19+18+17+16+15+14+13+12+11+10+9+8+7+6+5+4+3+2+1

Each pair, one from the line above and the corresponding one from the line below, now adds to 20 and there are 19 pairs, so that's 380. That is twice the total we want, so the answer is, again, 190.

Some readers may be thinking: 'There we go, there are 'tricks of the trade' at play here, and I would never have spotted that'. That's a consequence of the medium of the written word, where everything is inevitably tidied up. There is a story that, as a young boy, the mathematician Gauss was given by his teacher the task of adding up every number from one to one hundred. Gauss, a legendary prodigy, came up with the add-twice-and-halve shortcut, adding fuel to the myth that mathematicians are blessed with miraculous, instantaneous powers of insight. The reality is that, for most mathematicians, such insights incubate slowly and are the result of lots of playing around, leaving a problem, coming back to it and trusting that some insight will eventually emerge.

This playing around with effective and efficient ways for totting up the numbers from 1 to 19 brings us close to answering the second question of whether there is an easy way to find the total for any number of partygoers. How do we express what is going on here? The challenge now is, in John Mason's terms, to 'say what you can see', to articulate the move from the particular to the general. For 20 guests, we might say that we took 20 × 19 and halved it. That means for fifty guests we would take 50 × 49 and halve it. So for any number of guests, you multiply the number of guests by one less than that number and halve the answer. Algebraically, if the number of guests was $n$, then the number of handshakes is:

$$\frac{n \times (n-1)}{2}$$

## Re-presenting

We have reached a generalization, a rule that can be applied to any number, but is it really 'proved'? Although we may be convinced that the rule will work for any number, can we be absolutely sure? So far I have presented the problem purely through numbers and a little algebra. Another mathematical habit of mind is playing with different ways of presenting the mathematics, as different representations may help us to notice different things. Is there a more visual way to present our handshakes problem? Although the following image works with 20 people, it is easier to see what is going on by specializing a bit further, so let's look at a party of six.

We can imagine these six guests standing in a circle to shake hands and model this by placing six dots around the outside of the circle. Since everyone shakes hands with everyone else, we can join our dots to all the other dots (Figure 6.4):

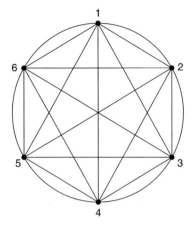

We can see from this that, not surprisingly, there are five lines from each dot (each of the six people shakes hands with five others). Since there are six dots, then 5 × 6 = 30 gives us the number of lines, yes? Well, no. Calculating 5 × 6 means we have counted each line twice; for example, the line joining 1 and 4 was counted in with Guest 1's five lines and again with Guest 4's five lines. So we have to halve that product: $6 \times 5 \div 2$.

Now imagine $n$ people standing in the circle. There will be $n - 1$ lines from each person, so the total number of lines is $n(n - 1)/2$; i.e. the same result as we got earlier.

Most people find this visual proof more convincing. The specific example of six people, presented in this fashion, makes the step to imagining what will happen with any number of people around the circle feel concrete in a way that playing around with numbers and algebra may not. The mark of a 'good' proof is not just that it logically establishes the veracity of something but also that it provides the reader with some insight into the situation, as this image of lines from points on the circumference does for many people. It grounds how the mathematics is 'working' and helps us see a structure behind the numbers and algebra. Although traditionally algebra is regarded as the bastion of proving, such visual proofs appeal in often being more intuitive, although for many centuries mathematicians were wary of the visual and would often hide their use of it, as we saw in the case of Gauss and the development of complex numbers.

# ▶ Proof by induction

So far, the approach has been grounded in proof by deduction. For a given number of guests, 20 or 200 or $n$, a logical argument is constructed for finding the total number of handshakes. The key thing is that each case stands alone, by which I mean that we could use the underlying structure to find how many handshakes there would be for 20 people without needing to know how many handshakes there would be for 19 or 21 people.

Playing around may not, however, so easily provide insight into the solution. So an alternative approach is to look at how the pattern builds up, from one person to two, and so on. If we can establish clearly the rule for building up from one number of guests to the next, then we are close to creating a proof by induction.

Proof by induction is commonly likened to a line of standing dominoes. To topple a line of standing dominoes in an orderly fashion relies on two things: first, that the dominoes are close enough to each other for each one as it topples to knock over the next one; and second, that the first domino gets knocked over. This is essentially the approach to using induction to prove the handshakes problem.

The starting point with induction is to specialize further, to go back to smaller cases and see if a rule for generating the pattern of numbers can be found. It helps to set out the results in a table.

| Number of guests | Number of handshakes | Total handshake1 |
| --- | --- | --- |
| 1 | 0 | 0 |
| 2 | 1 | 1 |
| 3 | 1 + 2 | 3 |
| 4 | 1 + 2 + 3 | 6 |
| 5 | 1 + 2 + 3 + 4 | 10 |

The pattern of totals is the triangular numbers that we met in Chapter 5. Saying what we can see, the fourth triangular number is the sum of the first four consecutive numbers. This was the 'seeing' that I used in the proof above. But another, building-up, 'seeing' is that the fourth triangular number is the third triangular number plus four. So the 20th triangular number is the 19th triangular number plus 20. In general terms, the $(n+1)^{th}$ triangular number is the $n^{th}$ triangular number plus $(n + 1)$.

The inductive reasoning step means that we now take as given the result for the $n^{th}$ triangular number and use that to create the $(n + 1)^{th}$ result. If we can show that this has the same form as the expression used for the $n^{th}$ when $n$ is replaced with $(n + 1)$ then, metaphorically, our dominoes are standing close enough together: if the expression is true for $n$, then it must always be true for $n + 1$.

As we saw above, when there are $n$ people at the party, there are $n(n - 1)/2$ handshakes (the $(n - 1)^{th}$ triangular number). So if there were $n + 1$ people at the party, the number of handshakes will be $n$ more, that is $n(n - 1)/2 + n$. We now need to reorganize this expression.

$$n(n - 1)/2 + n = n(n - 1)/2 + 2n/2$$
$$= (n(n - 1) + 2n)/2$$
$$= (n^2 - n + 2n)/2$$
$$= (n^2 + n)/2$$
$$= ((n + 1)n)/2$$

This is now the expression for the number of handshakes for $n + 1$ guests. To recap, what is established here is that if our expression for the total number of handshakes is correct for $n$, then it must also be correct for $n + 1$.

Now we have to check if our expression is correct for $n = 1$ (does the first domino get knocked over)? We know if there is one person at the party, then there are no handshakes. Putting $n = 1$ into our expression, we get $1(1 - 1)/2 = 1 \times 0/2 = 0$. Yes, the expression is correct for $n = 1$ and so our expression is correct for any number of guests.

This last step can look like a bit of sleight of hand, but it is not. We have shown that if our expression is correct for $n$, then it is correct for $n + 1$. It is correct for 1, so that means it is correct for 2. If it is correct for 2, then it is correct for 3, and so on, to an infinite number of guests at the party.

# ▶ Proof means looking at structure

Proving the handshakes problem involved looking for patterns and trying to predict what comes next. This is pattern sniffing, which is important to mathematicians. But important though pattern is, spotting a pattern alone is not enough. For example, what would you say comes next in these sequences:

1, 2, 4, 8, 16, ?

h, h, t, h, h, t, h, h, ?

People will generally agree that the answers are 32 and t. But those answers are only considered correct because those are the answers that everyone has agreed are correct. Patterns are not generated by the sequences of numbers or letters – the symbols themselves do not 'contain' anything; they only represent some other, extra-symbolic, situation. Take the second example. Suppose I tell you that is this the 'pattern' of heads and tails I got by tossing a coin eight times. Does that string of results mean that tails is going to come up on the next toss? No, because a coin does not have any pattern of results built into it. To really know what comes next in a pattern, we have to look at the underlying structure of what is generating the pattern. In the case of heads or tails, the structure is the probability of tossing a coin, and the 'pattern' is actually random – there is no predictive logic to it.

Taking the first example, before accepting an answer of 32, let us look at where this pattern might have come from. It relates back to the visual representation for the handshakes problem presented in Figure 6.4. Now, instead of counting the lines created by joining the dots on the circumference, we could count the number of regions created (Figure 6.5).

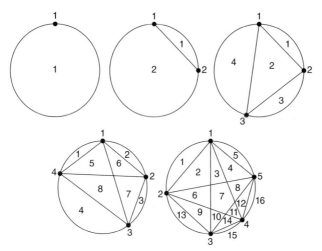

▲ Figure 6.5

We can see that one dot had one region, two dots two regions, three dots four regions, four dots eight regions and five dots 16 regions. Summarizing this in a table:

| Dots | Regions |
|---|---|
| 1 | 1 |
| 2 | 2 |
| 3 | 4 |
| 4 | 8 |
| 5 | 16 |

So how many regions for six dots? Obviously 32 regions. Actually, no; if you draw this and position the dots to maximize the number of regions, then the most you can ever get is 31 regions. Using this activity with students, inevitably some adamantly claim they have found 32 regions, or, having only found 31, keep redrawing the diagram in the hope of finding the missing region, which they won't. Unsettling as it is, the underlying structure here is not a doubling pattern, it only looks as if it is for the first few terms.

It is examples like this that drive the mathematician's scepticism and desire for proof. Here the seemingly obvious but incorrect pattern breaks down after only six steps; but who is to say that there are not some situations where what we think is the obvious pattern only becomes challenged after 600 or 6 million steps? Since it is impossible to ever check all the entries in a pattern, proof provides the mathematician with a test of certainty.

# ▶ Disproving is easier than proving

Given that, mathematically, for a conjecture to be proved, things have to be shown to be true in all possible cases, establishing that a conjecture is *not* true is an easier task. If only one example can be found that shows a conjecture is not true, then the conjecture has to be rejected or modified. Take, for example, this conjecture:

*The sum of any four even numbers is a multiple of four.*

We get a sense of what this conjecture claims by trying out some numbers. Take four even numbers – 2, 4, 8, 10. Their sum is 24, which is a multiple of four (4 × 6). Try another set of four – 2, 4, 6, 8. That sum is 20, another multiple of four. Seems reasonable; you are adding four numbers, each of which is even (divisible by two) and that feels like enough 'evenness' always to produce a multiple of four. How about 2 + 4 + 6 + 10 = 22? Oops, there goes that conjecture.

# Connecting

*Mathematics is the art of giving the
same name to different things.*

*Henri Poincaré*

Newborn babies, babies as young as three days old, are able to distinguish between faces. Given that the time such young babies actually have their eyes open amounts to only around 12 hours over the three days, this ability suggests that we are born with some sort of generic face recognition mechanism hard-wired into the brain. The way in which a young baby will lock eyes with its mother suggests that this mechanism involves, metaphorically at least, a pair-of-dots-pattern, and that scanning the environment for such a pattern is central to face recognition. As we grow older, we distinguish between faces instantaneously. While there is evidence that we are more attracted to reasonably symmetrical faces, it is unclear whether this is because a symmetrical face is a marker of good genes or is a reflection of having suffered less environmental stress. I say 'reasonably symmetrical' as we find perfectly symmetrical faces somewhat disconcerting – manipulate a digital photograph to make your own or a family member's face perfectly symmetrical and you will see what I mean.

## ▶ Symmetry

Our attraction to symmetry is not restricted to faces. As well as being drawn to the natural symmetry of vivid butterflies or fearful tigers, we build symmetry into objects, be they sandcastles or boardroom desks. Symmetry, natural and constructed, abounds, and we like it.

Considered in this way, symmetry appears to be a property of objects – a face either possesses symmetry or it does not. Mathematicians, in their habit of trying to understand the structures behind appearances, are interested in what symmetry does to objects (I'll explain what I mean by that later in this chapter), and this provides a good example of how mathematicians look for and find hidden connections between things that, on the surface, seem quite disparate.

At school we are taught about two types of symmetry: rotational and reflective. Let us explore these, both as a reminder of what they are and to set the scene for making some mathematical connections. Here's a simple thought experiment. Imagine, on a table, a small triangular piece of card, identical on each side and unmarked. If you leave the room and come back later, it is possible that triangle has been moved in some way and replaced but to all intents and purposes it looks the same. What are the different movements – (shape preserving) **transformations**, to use the mathematical term – which could have been applied to the triangle? For example, the triangle could have been picked up and rotated and put back down to look the same. Or it might have been flipped over and replaced. How many such different transformations are there?

Taking an image of a triangle, we can label the corners to keep track of things. There are two rotations that leave everything looking the same: a clockwise rotation through one-third of a full turn (120 degrees), and a clockwise rotation through two-thirds of a full turn (240 degrees) – see Figure 7.1.

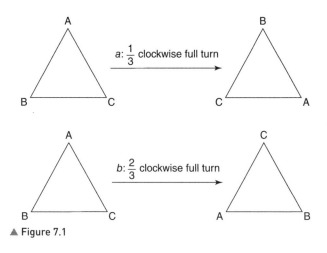

▲ Figure 7.1

Note that we could get the same results by different rotations. An anticlockwise rotation through two-thirds of a full turn positions the triangle exactly the same as one-third of a full turn clockwise, or as does four one-thirds of full turns clockwise. Every instance where the triangle ends up in the same position is equivalent to a simple rotation, which I am going to take as 120 degrees and 240 degrees clockwise. For short, let's call these *a* and *b*, respectively.

The triangle can also be 'flipped' (reflected) along the direction of a line from each corner to the centre of the opposite side. Let's call these *r*, *s* and *t* (see Figure 7.2).

Figures 7.1 and 7.2 show five things that could be done to the triangle in your absence and you would be none the wiser. And there is one other thing that could have

*Connecting*

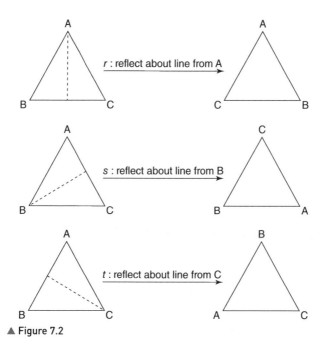

▲ Figure 7.2

happened – that the triangle was left alone, was not moved at all. Although nothing has changed, mathematicians call this the **identity transformation** and usually denote it *e*. It is not clear why the mathematical convention is to use *e* instead of *i* here, but it is reasonabe to suppose that it is to avoid confusion with the imaginary number $\sqrt{-1}$.

Now that we have these six transformations as our building blocks, we can look at what happens when you combine them, one after another. Suppose you did *a* followed by *r*, that results in Figure 7.3.

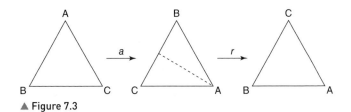

▲ Figure 7.3

We could, however, have got to that final position directly by carrying out *s* alone. Is it the case that any pair of these transformations has the same effect as a single transformation? It seems sensible to expect that they should; after all, we have only got six different positions that the triangle can placed in, so it would be very strange if some new position suddenly became possible. Playing around with cut-out triangles reveals that this is what happens: apply any one of these transformations followed by a second and you can find a single transformation that has the same effect.

Such observations prompt the mathematician to want to express this concisely. Rather than continuing to draw triangles, some new notation helps. We have already used letters to represent the transformations; a symbol now allows us to express combinations. It is conventional to use a small circle to denote following one transformation with another, and to record the order from *right* to *left*. So transformation *a* followed by *r* would be:

$r \circ a$ (apply rotation *a* first, then reflection *r*)

As this combination has the same result as the transformation *s*, it makes sense to complete the equation:

$r \circ a = s$

The mathematically curious now want to know what the complete set of all pairwise combinations of transformations looks like. As there are six transformations, there are 6 × 6 possible pairings and an effective way to look at all of these is to set them out in a table. Convention demands that the left-hand column designates which transformation to do first, the top row the second. Filling in $r \circ a = s$ gives us the result in Figure 7.4.

| ○ | e | a | b | r | s | t |
|---|---|---|---|---|---|---|
| e |   |   |   |   |   |   |
| a |   |   |   | s |   |   |
| b |   |   |   |   |   |   |
| r |   |   |   |   |   |   |
| s |   |   |   |   |   |   |
| t |   |   |   |   |   |   |

▲ Figure 7.4

We can save time filling in the other 35 entries by dealing with the identity transformation, e. Whether we do e first or second, the effect of combining it with another transformation is the same as doing that transformation on its own, so the first column and top row are easily filled in (Figure 7.5).

| ∘ | **e** | **a** | **b** | **r** | **s** | **t** |
|---|---|---|---|---|---|---|
| **e** | e | a | b | r | s | t |
| **a** | a | | | s | | |
| **b** | b | | | | | |
| **r** | r | | | | | |
| **s** | s | | | | | |
| **t** | t | | | | | |

▲ Figure 7.5

We might also notice that the reflections 'undo' themselves; r followed by r, for example, has the same effect as e. Similarly, a or b followed by the other has the same effect as e (Figure 7.6).

| ∘ | **e** | **a** | **b** | **r** | **s** | **t** |
|---|---|---|---|---|---|---|
| **e** | e | a | b | r | s | t |
| **a** | a | | e | s | | |
| **b** | b | e | | | | |
| **r** | r | | | e | | |
| **s** | s | | | | e | |
| **t** | t | | | | | e |

▲ Figure 7.6

Filling in the remaining entries is not difficult (Figure 7.7).

| ∘ | **e** | **a** | **b** | **r** | **s** | **t** |
|---|---|---|---|---|---|---|
| **e** | e | a | b | r | s | t |
| **a** | a | b | e | s | t | r |
| **b** | b | e | a | t | r | s |
| **r** | r | s | t | e | b | a |
| **s** | s | t | r | a | e | b |
| **t** | t | r | s | b | a | e |

▲ Figure 7.7

# ▶ Patterns in the table

Now we can examine the table for patterns. The table confirms what we expected in that there are no surprise results; every element in the table is one of the original transformations. In mathematical terms, we have closure. One thing you might notice is that e appears in every row and every column. That tells us that each of these transformations has an inverse; for example, the inverse of a is b since a ∘ b = e. As already noted, we can see that the three reflections are self-inverses; r ∘ r = e and so forth.

One other result, not immediately obvious but easily checked, is that if we carry out three consecutive transformations, the result is the same no matter which pair we combine first, providing the order of the transformations is not changed. In other words, taking $a \circ b \circ r$, whether we do $a \circ (b \circ r)$ or $(a \circ b) \circ r$ the result will be the same.

And the point of this? These symmetries of the equilateral triangle are an example of what mathematicians call a **group**. A mathematical group is a set of elements and an operation between the elements that has four properties.

▶ Closure: combining one element from the group always results in another element of the group.

▶ Identity element: there is a member of the group that when combined with any other element from the group results in that same element.

▶ Inverses: for every element in the group, there is another element such that when these two elements are combined, the result is the identity element.

▶ Associativity: the result of several consecutive operations is the same regardless of the order of pairing, providing the order of the operations is not changed.

# ▶ Connections

Setting out the structure of the triangle symmetry group as shown above is known as putting it into a Cayley table, named after the 19th-century British mathematician Arthur Cayley. This arrangement is similar to the tables of addition or multiplication facts that we meet at school. A Cayley table can help make clear many of the properties of the elements and their combinations. The addition table usually deals only with the natural numbers, but we can extend this to the integers (Figure 7.8). As these are infinite, we can only record part of the Cayley table but that is sufficient to look for and think about patterns and properties.

| + | −3 | −2 | −1 | 0 | 1 | 2 | 3 |
|---|----|----|----|---|---|---|---|
| **−3** | −6 | −5 | −4 | −3 | −2 | −1 | 0 |
| **−2** | −5 | −4 | −3 | −2 | −1 | 0 | 1 |
| **−1** | −4 | −3 | −2 | −1 | 0 | 1 | 2 |
| **0** | −3 | −2 | −1 | 0 | 1 | 2 | 3 |
| **1** | −2 | −1 | 0 | 1 | 2 | 3 | 4 |
| **2** | −1 | 0 | 1 | 2 | 3 | 4 | 5 |
| **3** | 0 | 1 | 2 | 3 | 4 | 5 | 6 |

▲ Figure 7.8

Are any of our conditions for a group satisfied?

▶ Addition of integers is closed: add any pair of integers and the answer is always an integer (even if we have to imagine our table being extended to show the answer).

▶ Is there an identity element? Yes; zero added to any integer leaves it unchanged.

▶ Does every integer have an inverse under addition? Yes; the positive and negative pairings of integers join up as inverses. For example, the inverse of 3 is –3 as 3 + (–3) = 0 (and likewise the inverse of –3 is 3). That every element has its inverse is also evident from 0 appearing in each row and column of the table.

▶ Finally, is addition associative? Yes; as we saw earlier, adding three numbers we get the same result irrespective of which pair we add first, for example (5 + (–3)) + 6 = 5 + ((–3) + 6).

So the set of integers under the operation of addition forms a group. Although one obvious difference between the group of integers under addition and the group of symmetries of the equilateral triangle is that there are an infinite number of elements in former, there is another key difference. Imagine a line down the diagonal of the addition Cayley table from the top left-hand corner to the bottom right-hand corner. This acts as a mirror line, with the upper right half of the table mirroring the bottom left half. This is not surprising, as we know addition is commutative: 5 + 7 = 7 + 5. The Cayley table for the symmetries of the equilateral triangle does not, however, have this property. Whereas the order of adding two integers does not matter in the sense of affecting the final answer, the order of carrying out two transformations on the triangle does matter. Groups with this additional property of commutativity are called Abelian, named after the Norwegian mathematician Niels Henrik Abel.

# What about multiplication?

If addition of integers forms a group it would seem likely that multiplication would do too. Again, a Cayley table helps us to explore whether this is the case (Figure 7.9).

| ×  | −3 | −2 | −1 | 0 | 1  | 2  | 3  |
|----|----|----|----|---|----|----|----|
| −3 | 9  | 6  | 3  | 0 | −3 | −6 | −9 |
| −2 | 6  | 4  | 2  | 0 | −2 | −4 | −6 |
| −1 | 3  | 2  | 1  | 0 | −1 | −2 | −3 |
| 0  | 0  | 0  | 0  | 0 | 0  | 0  | 0  |
| 1  | −3 | −2 | −1 | 0 | 1  | 2  | 3  |
| 2  | −6 | −4 | −2 | 0 | 2  | 4  | 6  |
| 3  | −9 | −6 | −3 | 0 | 3  | 6  | 9  |

▲ Figure 7.9

As for addition, the condition of closure is met: multiplying any two integers together always results in another integer. There is an identity element: 1. And multiplication is associative: $(a \times b) \times c = a \times (b \times c)$. But the integers do not all have multiplicative inverses. Taking, for example, 3: there is no integer, $p$, that makes $3 \times p = 1$ true (1 will never appear as an entry in either the row or the column headed by 3). Unlike addition of integers, multiplication of integers does *not* form a group. Thus there are fundamental differences in the underlying structures of integers under addition and multiplication, differences that group theory allows mathematicians to explore.

# ▶ Wallpaper groups

Any trip to a decorating store can make one feel overwhelmed with the plethora of books of wallpaper designs. It seems that there must be a limitless set of possibilities, which of course there are, as designers can create a myriad different repeating motifs. Assuming that any motif is not changed in size, all wallpaper designs are created out of four basic transformations:

▶ reflections

▶ rotations (as with the symmetries of the equilateral triangle)

- translations (a sliding move)
- the hybrid glide reflection that combines a reflection and a translation.

Group theory shows that if we strip away the different motifs, given these basic transformations there are, in essence, only 17 different wallpaper designs.

The most famous collection of these 17 designs is in the Alhambra palace in Spain, although they are preserved in tiles, not paper. The 13th-century Moorish artists who created these tilings not only created something of stunning beauty, but they did it all without measuring angles or lengths, the only available tools being a pair of drawing compasses and a straight-edge (like a ruler but with no markings on it). Or so it used to be assumed. There is some recent evidence that like the modern-day wallpaper designer, these artists may have worked with templates of motifs, which they used to mark out the designs on plaster. However the mosaics were created, they embody the mathematics of group theory, a theory that would not be apparent to mathematicians until over 500 years later.

The significance of these wallpaper symmetry groups extends beyond the aesthetics and is central to the study of crystals. The 17-fold classification was in fact discovered and classified by a crystallographer, the Russian E. S. Federov, and group theory continues to play a key role in crystallography.

# ▶ Monster group

Not content with the likes of wallpapers and tiles, the mathematician's search for ever-extending generality led the North American mathematician Robert Griess in 1981 to what is known as the **monster group**. Beyond imagination, a metaphor for this group is that it is a mathematical snowflake with $10^{53}$ symmetries. As if that were not mind-boggling enough, it exists in a 196 884 dimensional space.

Is the monster group simply a mathematical freak show? The mathematics of this group already connects symmetry with physics, and it appears to have the potential to provide further insights into string theory. Once again, the mathematics of the playful, the abstract, the seemingly disconnected from reality finds connections back into the physical world.

# Modelling

*The knowledge of which geometry aims is the knowledge of the eternal.*

*Plato*

Mathematical problem-solving, when directed at solving real world problems as opposed to purely mathematical exploration, begins with creating mathematical models of the world. Setting up mathematical models is something we all do with ease from an early age. Figuring out how many chocolates have been eaten from a box of ten if six remain, the child putting up six fingers and counting how many more fingers she needs to raise to make ten is setting up a mathematical model where fingers model chocolates. Although the problems may change and become more complicated and the range of available mathematical tools expanded, in essence all mathematical modelling is analogous to this example.

# ▶ The modelling cycle

A diagram, a model of modelling (Figure 8.1), illustrates the relationship between the real, physical world and the mathematical world.

Another simple example illustrates the movement around this cycle and in particular the important step of

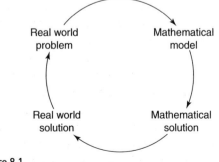

▲ Figure 8.1

moving back into considering the real world implications of any mathematical solution. Suppose a school of 1650 students is going on a field trip and the coaches for hire can each carry 68 passengers. How many coaches will need to be hired?

Secondary school students easily set up 1650 ÷ 68 as a mathematical model of the situation. Punching this into a calculator or doing a paper and pencil calculation gets the mathematical answer of 24.265 or 24 with remainder 18. What is perhaps surprising is the number of students answering that 24.265 coaches need to be booked, not pausing to wonder what 0.265 of a coach would look like. Or that 24 coaches would suffice without realizing that this would leave 18 people behind. They skip over the vital last stage of the modelling cycle of converting a mathematical answer into a sensible real-world solution.

For mathematicians, this process of setting up models goes beyond simply using models to get answers for real-world problems. They are also interested in studying the models themselves, playing with and extending them. As long as such extensions preserve the internal consistency of the mathematics, then whether or not these extended models represent anything back in the real world becomes irrelevant, although one of the remarkable features of mathematics is how often seemingly completely abstract models do end up having real-world explanatory power. The move from our physical world of two and three dimensions to higher dimensions illustrates this process and the story begins with Cartesian coordinates, invented by the French mathematician and philosopher René Descartes.

# ▶ Cartesian coordinates

Accounts of how Descartes developed the idea suggest that it came to him either in a dream or while watching a fly above his counterpane: whichever was the case, he seems to have found his inspiration in bed. Descartes' insight was that he could describe the position of the fly with precisely three numbers. Initially tucked away in an appendix to his famous work *The Discourse on Method*, its impact on mathematics is huge.

Cartesian coordinates in the two-dimensional plane use the familiar $x$ and $y$ axes. A point is chosen somewhere in the plane to be the origin (0, 0), a unit of measurement is used to set up perpendicular axes, and the position of points from the origin are recorded as, conventionally, the horizontal distance from the origin followed by the vertical distance. So in Figure 8.2, (5, 1) is a different point from (1, 5).

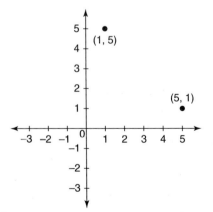

▲ Figure 8.2

The use of coordinates now is ubiquitous (satnav systems would be impossible without them) and Descartes' invention of them changed mathematics forever. Prior to this, geometry and algebra were essentially distinct branches of mathematics. Geometry (literally 'earth (geo) measurement') was focused primarily on measuring continuous quantities (water, land area, mass of corn) and studying the properties of shapes. Algebra had grown out of counting discrete quantities (fish, trees, bushels) and calculating. The invention of Cartesian coordinates meant that geometric forms could be expressed algebraically, and algebraic ideas, which it had previously not been possible to visualize, could be given form through geometrical embodiments. Geometry, released from relying on diagrams, could be handled analytically. And new geometries, beyond our embodied world of three dimensions, could be brought into being.

# ▶ The distance between two points

Suppose that in the two-dimensional plane there are two points, *P* and *Q*, and we need to find the distance between them. Prior to the invention of Cartesian coordinates, we would have had to connect up the points and measure the length of the line segment. Coordinates provide another way of finding the distance. Setting up the coordinate system point, suppose point *P* turns out to be located at (2, 1) and point *Q* at (5, 5), as in Figure 8.3.

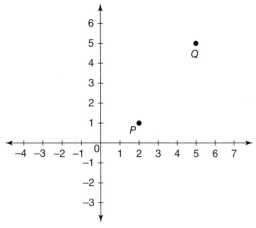

▲ Figure 8.3

This is the first move in the modelling cycle, from the real world to a mathematical model of the essential features needed to be used in solving the problem. Like any model, non-essential features are not included. We do not need to know if $P$ and $Q$ mark the positions of buried treasure, whether or not travelling between them is possible, or if they are on land or in the sea. The situation is reduced to its bare mathematical minimum.

Now we can work on the mathematical model. Constructing a right-angled triangle (see Figure 8.4), the coordinates of the right angle $R$ can be found and the lengths of the horizontal and vertical sides calculated. Pythagoras' theorem allows us to calculate the length of the hypotenuse, the distance between our two points.

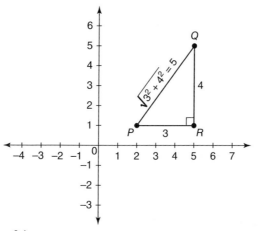

▲ Figure 8.4

Playing around with more examples, it may become clear that we can make this model more abstract and calculate the missing length without having to draw the coordinates. If our two points have coordinates $(a_1, b_1)$ and $(a_2, b_2)$, the lengths of the horizontal and vertical sides that could be imagined as constructed are the difference between $a_1$ and $a_2$ and the difference between $b_1$ and $b_2$. Writing $(a_1 - a_2)$ and $(b_1 - b_2)$ for these lengths might initially look worrying because, what if $a_1$ was 5 and $a_2$ was 7 then $(a_1 - a_2) = -2$. A negative distance? Looking ahead, applying Pythagoras' theorem means that this value is going to be squared, and since squaring a negative number results in a positive value, we can go ahead and use $(a_1 - a_2)$ and $(b_1 - b_2)$ irrespective of whether these values are positive or negative. Putting these values into the Pythagorean formula we get that the distance between the two points is:

$$\sqrt{(a_1 - a_2)^2 + (b_1 - b_2)^2}$$

Descartes realized that a third coordinate lifts a point up out of the plane and can specify its position in three-dimensional space. If our two points are then $(a_1, b_1, c_1)$ and $(a_2, b_2, c_2)$ the formula for calculating the distance between two points in three dimensions generalizes from the formula for two dimensions to:

$$\sqrt{(a_1 - a_2)^2 + (b_1 - b_2)^2 + (c_1 - c_2)^2}$$

## Fermatian coordinates?

Around the time that Descartes was developing coordinates, Pierre de Fermat, a French lawyer and amateur mathematician, was also playing with developing a coordinate system, one based on the axes not being at right angles to each other. Although Descartes bagged the prize for coordinates, Fermat earned his place in the mathematics hall of fame for his last theory. From Pythagoras' theorem we know that there are infinitely many solutions to $x^2 + y^2 = z^2$ with integer values for $x$, $y$, $z$, for example, 3, 4, 5 or 5, 12, 13 (known as, not surprisingly, Pythagorean Triples). Fermat's last theorem is that 2 is the highest power to which three numbers can be raised and for there to be integer solutions for the sum of two of these to equal the third. In other words, there are no integer solutions to $x^3 + y^3 = z^3$, or $x^n + y^n = z^n$ where $n$ is an integer. Strictly speaking this should be known as Fermat's Last Conjecture or Hypothesis as he left no proof – the claim was written in the margin of a book with a note from Fermat that he had a proof but could not fit it in the margin. It took mathematicians over 350 years to finally come up with a proof, which the British mathematician Sir Andrew Wiles released in 1994.

# ▶ Generalizing the model

Cartesian coordinates allow us to specify the position of mathematical objects, for example squares or cubes. Just as scientists setting up experiments like to control things by reducing the number of variables, so do mathematicians. The unit square and unit cube – shapes where all the sides have length 1 – control the variable of lengths of sides. Life is also made simpler by setting up these shapes and the coordinate axes so a vertex of the square or cube sits at the origin (see Figure 8.5).

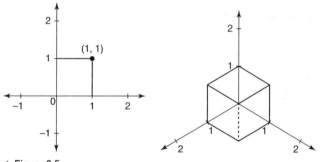

▲ Figure 8.5

From this we can note down the coordinates of all the vertices of each shape:

Square: (0, 0), (0, 1), (1, 0), (1, 1)

Cube: (0, 0, 0), (0, 0, 1), (0, 1, 0), (1, 0, 0), (1, 1, 0), (1, 0, 1), (0, 1, 1), (1, 1, 1)

Are there patterns to sniff out here? Obviously, given how we set things up, the coordinates only involve 0 and 1, but, importantly, they list all possible permutations of these two values, taken in pairs or threes. The vertices of the square are the four possible permutations of taking pairs of 0 and 1; the vertices of the cube are the eight permutations of taking triples of 0 and 1.

But why stop there? We could take an ordered set of four 'coordinates' (a, b, c, d). If each value is restricted to 0 or 1, then there are 16 possible permutations from (0, 0, 0, 0), (0, 1, 0, 0) through to (1, 1, 1, 0), (1, 1, 1, 1). These are the 16 coordinates of the four-dimensional hyper-cube.

Every time another 'coordinate' is added, the number of permutations of 0 and 1 doubles, so in five dimensions the unit shape has 32 coordinates. There is nothing to stop us doing this for any whole number n. And so n-dimensional space is born.

# A five-dimensional coordinate is a coordinate of what?

In everyday talk, we use the word 'abstract' to mean at least two different things. There is the sense of pulling out the essential features, akin to extracting, as when one of Picasso's abstract portraits is not at all life-like but still recognizably a person. Then there is abstract in the sense of not being even remotely linked to anything in real-life – a Jackson Pollack painting comes to mind. Mathematicians play with both these senses of abstract. Starting from the real world of stakes in the ground or cubes of wood, some essential features are 'abstracted out' and represented in a mathematical model, the coordinate system in this case. The mathematical model then becomes explored in its own right, and a further abstraction made, generalizing from coordinates with two or three ordered numbers to 'coordinates' with four, five – or however many you like – ordered numbers. I put quotation marks around the word 'coordinates' in the last sentence to flag up this subtle but important shift. When I read the coordinates of a point off a map, my interest is in the point and the map (Am I lost?) and my attention is on the real world, with the map helping me to navigate around the world. In moving to four or n dimensions, the mathematician's attention is on the mathematical world, and the interest is in the properties and patterns of the coordinates as objects in their own right, not on what those coordinates might represent. A coordinate like (0, 1, 1) straddles these two worlds of the real and the mathematical – it can represent something out there in the physical world; equally, it can be a purely mathematical object. Asking 'A five-dimensional coordinate is a coordinate of what?' is based in the first use of the abstract – that there has to have been something in the real world that such a coordinate represents. To the mathematician, the question is not meaningful as five-dimensional coordinates are not representing anything – they are abstracted from mathematics, not from the world. They are, however, meaningful (i.e. full of meaning) in the sense of being consistent with the rules of the mathematical world as established through the properties of two- and three-dimensional coordinates. For example, we saw in Figure 8.4 that calculating distance in the two- and three-dimensional plane was done using particular formulae. Again, we can generalize. So, for example, taking two points in five-dimensional space, we can define the distance between these two points as:

$$\sqrt{(a_1 - a_2)^2 + (b_1 - b_2)^2 + (c_1 - c_2)^2 + (d_1 - d_2)^2 + (e_1 - e_2)^2}$$

Distance here is metaphorical; it is not a literal distance, but calling this abstract quantity 'distance' preserves the echoes of its real-world origins, giving us some handle on what would otherwise be such an abstract concept that working with it would become reduced to a meaningless manipulation of symbols. I cannot literally imagine a five-dimensional distance, but it has some meaning through its metaphorical linking to actual dimensions.

These activities of playing with and extending models, of abstracting to the point of abstraction, might have restricted mathematics to simply an intellectual game were it not for mathematics' 'unreasonable effectiveness', in the words of the Nobel Prize-winning Hungarian physicist Eugene Wigner. Mathematics is unreasonably effective because of the myriad ways in which purely mathematical ideas turn out to be capable of being applied back into the real world and so help us understand its secrets. The idea of four dimensions was brought to life by Einstein binding time to space – could this imagining of the space-time continuum have been possible without the abstract mathematics of $n$-dimensions?

For many, even those not familiar with the details of Einstein's theory, the sense of a four-dimensional world has become commonplace. **String theory**, involving even higher dimensions, is still on the edge of most people's understanding. Again, the naming is metaphorical – string theory is an extension of the properties of minute strings vibrating. Theorists in string theory happily work in worlds with 10 and/or 11 dimensions. And/or? Surely it must be one or the other? For the mathematics of string theory to be consistent, the number of dimensions of space-time can be either 10 or 11 depending on different starting assumptions. Can we begin to imagine what the other six or seven dimensions might be? Perhaps not, but as mathematicians develop more models (of models) and metaphors, to future generations this may be as obvious as a three-dimensional world is to us.

# ▶ The Koch snowflake

Take any equilateral triangle, mark off each side into thirds, construct equilateral triangles on the outside of each middle third and you create a six-pointed star. That star has 12 edges; repeating this process of 'third-and-construct' on each of these 12 edges creates a shape with 18 vertices. Continuing indefinitely on each new edge produces a snowflake of sorts, one with increasingly wrinkly points (see Figure 8.6).

Named after the Swedish mathematician Helge von Koch, who presented it in a 1904 paper, the Koch snowflake is not only a beautiful object, but also has some interesting properties. Imagine enclosing the snowflake snuggly within a circle. No matter how many iterations of third-and-construct we carry out, the snowflake will never break free of its circular boundary. We may not be able to calculate the area of the Koch

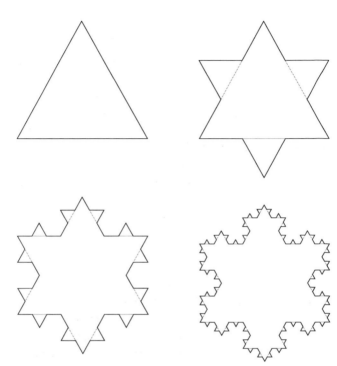

▲ Figure 8.6: The Koch snowflake

snowflake, but we know for certain that its area is going to be less than the area of the circle.

The area of the Koch snowflake is finite.

As the perimeter of the Koch snowflake becomes more and more crinkly, so it becomes longer and longer. The more we zoom in on the perimeter to try to measure the length, the more kinks we find, adding more and more length to the perimeter.

The perimeter of the Koch snowflake is infinite.

# ▶ Distance revisited

Earlier in this chapter, in setting up the problem of the distance between two points, I blithely suggested that the model could ignore certain features such as where the points were in the world and whether you could travel between them – a straight line joining them sufficed as a measure of the distance between them. Lots of mathematics, and geometry in particular, makes such sweeping assumptions, often because it makes the resulting mathematics easier to deal with. These assumptions meant I could use Pythagoras' theorem to calculate the length of the line – I could not do that with a curved line joining the points.

But suppose we want to be truer to the real world? In the world, the distance between two points is rarely a straight line. The relatively obscure British mathematician Lewis Richardson was concerned with this problem of measuring distances more accurately. His initial interest in the likelihood of adjacent countries going to war led him to investigate whether this was linked to the length of any common border. Gathering data on various countries, Richardson was struck by the differences in the reported lengths of shared borders. Depending on the source, the border between Spain and Portugal was reported as either 987 km or 1214 km. Interested now in the challenge of measuring borders and edges, in the 1930s Richardson published a paper called 'How long is the coastline of Britain?' introducing what has come to be known as the 'coastline paradox'.

Imagine that you have some impossibly long ruler with which to measure, a ruler 100 km long. Measuring the coastline of Britain by laying the ruler down so that both ends touch the coast is clearly going to lead to a measurement that is wildly inaccurate. Snapping the ruler in half and using a 50 km ruler is going to be a bit more accurate and will also result in reporting the coastline as longer. Continuing to halve the length of the ruler produces increasingly accurate measures that are also increasingly longer. Richardson argued that this process leads to the conclusion that the length of the coastline increases without limit.

The coastline of Britain is infinite.

# ▶ Fractals

When published, Richardson's work did not create many ripples in the mathematics community. Likewise, the Koch snowflake was something of a mathematical curiosity. Then the Polish-born American mathematician Benoit Mandelbrot brought these two ideas together. The key connection between the real world of coastlines and the mathematical world of perfect snowflakes was encapsulated in Mandelbrot's work on **self-similarity** – the idea that part of an object is similar to the whole.

Self-similarity is a property that classic Euclidean geometric shapes do not have: zoom in on part of the circumference of a circle and it looks nothing like a circle. Zoom in, however, on part of a Koch snowflake and it looks like a Koch snowflake. Similarly, the coast of Britain photographed from a satellite looks similar to the outline of the coastline photographed from a helicopter. Mandelbrot coined the term 'fractal' to describe objects with this property of self-similarity. It turns out that the physical world is much more 'fractal' than it is Euclidean: blood vessels, ferns, broccoli are all self-similar. Apart from creating beautiful computer-generated images, fractals are opening up new adventures in geometry and raising further questions about dimension.

In the discussion of $n$-dimensions, the generalization rested on n being a whole number. It does not make sense to talk about creating a coordinate with, for example, 2.3 numbers. Yet study of the dimensions of fractals shows that it is possible to think of objects as having a dimension that lies between whole numbers. Intuitively an object has a fractal dimension when it is too complex to be one-dimensional but too simple to be two-dimensional, as is the case with Koch snowflakes and coastlines. Once again, mathematics is proving its 'unreasonable effectiveness'.

# Puzzling

*The infinite! No other question has ever moved so profoundly the spirit of man.*

*David Hilbert*

*There are two misstakes in this sentence.*

Is the sentence above true or false? At first glance it looks as though there is only one mistake – the misspelling of 'mistake'. But that would mean the claim that there are two mistakes in the sentence is also a mistake. So if there are two mistakes, the sentence is true. But if it is true, then the claim that there are two mistakes is not a mistake, so it is false … It's like that schoolyard prank of handing someone a piece of paper reading 'How do you keep a fool busy? Answer on reverse.' Turning the paper over, you read the same thing again.

We can never decide whether or not the sentence is true or false; it is neither. For a subject like mathematics, built up on the idea of starting from statements – axioms – that could be self-evidently taken as true, the emergence of the idea that some statements could be undecidable, neither true nor false, was a threat to the whole foundation of the discipline.

# ▶ The barber paradox

Bertrand Russell is acknowledged to have been the first to bring this tension to the attention of the mathematics community. His version of this undecidable paradox was couched in the metaphor of a barber.

In a certain town there is one male barber who shaves every man who does not shave himself. Who shaves the barber?

If the barber shaves himself then he cannot shave himself, but if he does not shave himself then he should shave himself. Perhaps the sensible response to this conundrum is simply to say that no such barber could exist, but it prompted Russell to work on refining a branch of mathematics known as **set theory** and for a while it seemed that mathematics was back securely resting on notions of true or false.

Such paradoxes, however, were further explored by Kurt Gödel, who challenged the dichotomous true/false view, developing the argument that rather than the basic axioms of arithmetic being indisputably true or false, they could lead to contradictions, that the logic of mathematics was inevitably 'incomplete' (hence Gödel's 'incompleteness theory'). This theory crushed the holy grail of mathematics, sought for centuries, that there was an ultimate set of unquestionably 'true' axioms upon which the whole of mathematics could be constructed.

Challenging though Gödel's work was to the mathematics community, rather than marking the collapse of mathematics it opened up new possibilities in mathematical applications, particularly in computing, physics and economics. Gödel was a friend and colleague of Albert Einstein at Princeton University and it is likely that the ideas of the theory of relativity would have been unthinkable without this work of Gödel.

Gödel's work also opened up space for a new branch of mathematics known as **fuzzy logic**. Whereas classical logic is based on conditions being true or false, fuzzy logic, initially developed by the computer scientist and mathematician Lotfi Zadeh, is based on a *range* of truth-values. Zadeh applied this logic to applications like thermostats, but the community of engineers was initially resistant to Zadeh's ideas – they brought an unsettling sense of vagueness to the field. Such a reaction is hardly surprising – who wants to drive over a bridge built on fuzzy logic? But like so many mathematical ideas that at first seem threatening to the status quo, many things that we now take for granted – self-focusing cameras, 'smart' washing machines, anti-lock braking systems to name but three – are controlled by fuzzy logic.

# ▶ Counting the infinite

Infinity is the mathematician's sandbox. Strange things go on at infinity, but we do not have to travel that far to get of sense of why. Take the following question: 'Which are there more of, the natural numbers (1, 2, 3, …) or the even natural numbers (2, 4, 6, …)?'

An intuitive answer is that the collection (**set**) of even numbers must be half the size of the first set. If we think of the counting numbers as represented by a bag (infinitely large) of marbles, each labelled with one of the natural numbers, then removing all the even-numbered marbles would leave only the odd numbered marbles, a set of marbles the same size as the set removed, thus halving the total number that was in the bag originally.

But imagine two bags of marbles, one (Bag A) containing all the natural numbers, and the other (Bag B) only containing the even natural numbers. We can imagine taking the even numbered marbles out of Bag B one at a time; we can count these by matching them up – putting them in one-to-one correspondence – with the marbles from Bag A: 1 matched to 2, 2 to 4, 3 to 6 and so forth. Neither Bag A nor Bag B is going to run out of marbles before the other does: there are the same number of infinite even numbers as there are infinite natural numbers.

Counting in this way is no different from the counting of young children. Given a collection of six toys to count, the child has to coordinate three one-to-one correspondences – matching one object to one point to one number word. Skipping an object, pointing to an object twice or getting the order of numbers wrong are errors all youngsters make in figuring out how to count a collection, but by the age of five, most of us can do this accurately and without difficulty. In theory, we then know that we could count any sized collection of objects, although in practice with very large collections we are content to work with approximate numbers.

The playful mathematician takes this simple understanding of counting as making one-to-one correspondences and asks: 'What would happen if we applied this approach to infinite collections, collections that we could never actually completely count? What would be the implications?' As we saw above, infinity does not behave in the same way that finite numbers behave. If I have a bag of 1000 numbered marbles, then exactly 500 of them are the even-numbered ones, but the logic of 'there are half as many even numbers as there are counting numbers' ceases to hold at infinity.

'At infinity' is itself a nonsense phrase – it implies that infinity is a place that, if only given enough time, we could reach: infinity is just a very large number. But as the thought experiment about counting even numbers reveals, we have to think differently about infinity.

## ▶ What else is countable?

I have argued that counting the infinite is grounded in the everyday activity of counting by putting objects into one-to-one correspondence with each other. Since I can, in my imagination, set out the even numbers in an never-ending line, I can imagine counting them. This same logic can be applied to counting other collections of numbers. Galileo expressed a similar paradox in thinking about the square numbers: 1, 4, 9, 25, … This pattern continues infinitely, but can, just like the even numbers, be counted.

What about counting the extensions to the number system that mathematicians invented? All those fractions – the rational numbers – can they be counted? To be able to count all the rational numbers we need to lay them out in a row so that we can put the counting numbers in one-to-one correspondence with them. We saw in Chapter 4 that the number line is tightly packed with rational numbers: given any two fractions, no matter how close together, we can always zoom in and squeeze another fraction between them. But if every pair of rational numbers in our row can have another put between them, we can never get them set out in a way that allows us to begin to count them.

Except we can.

The brilliant insight of the mathematician Georg Cantor was to lay out the rational numbers not in a row but in an array. The top row of the array has all the rational numbers with denominator one, the second row all fractions with denominator two, the third row with three, and so forth, continuing infinitely. The first column then has all the rational numbers with numerator one, the second column with numerator two, and so on, again extending infinitely. (We could interweave into this array all the negative rational numbers by alternating the columns with positive and negative, but here I am just going to show the argument for counting the positive rational numbers – the argument proceeds in a similar fashion with the negative ones included). The beginning of the array looks like Figure 9.1.

$$\frac{1}{1} \quad \frac{2}{1} \quad \frac{3}{1} \quad \frac{4}{1} \quad \frac{5}{1} \quad \ldots$$

$$\frac{1}{2} \quad \frac{2}{2} \quad \frac{3}{2} \quad \frac{4}{2} \quad \frac{5}{2} \quad \ldots$$

$$\frac{1}{3} \quad \frac{2}{3} \quad \frac{3}{3} \quad \frac{4}{3} \quad \frac{5}{3} \quad \ldots$$

$$\frac{1}{4} \quad \frac{2}{4} \quad \frac{3}{4} \quad \frac{4}{4} \quad \frac{5}{4} \quad \ldots$$

$$\frac{1}{5} \quad \vdots \quad \vdots \quad \vdots \quad \vdots$$

▲ Figure 9.1

Does our array contain all the possible rational numbers? Yes; given any fraction, for example 352/456 or 768/42, we can find it in the array by going down to the row containing that denominator and along to the column with that numerator. The array also contains equivalent fractions: for example, 1/2 is there, and so are 2/4, 3/6 and every other possible equivalence to 1/2 (of which there are an infinite number!). We could go through and remove all these duplicates, but it does not make any difference – if we can count every entry in this array, then we can certainly count every fraction in the array with the duplicates removed. What is important that we are convinced that every possible rational number is somewhere in this array, that this net catches them all.

Having thus laid out the entire set of rational numbers, we can now count them. It is no good starting by counting the top row as we will never get to the end of it and the second row will never get counted. The system to counting every element in the array is to follow the arrows in Figure 9.2.

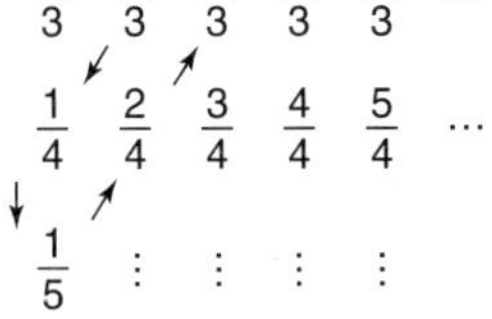

▲ Figure 9.2

Following this route ensures that we visit and count each and every rational number. As before we won't run out of whole numbers to count with before we reach the end of the rational numbers – we can match the counting numbers in one-to-one correspondence with the rational numbers. The rational numbers is a countable set of numbers.

# ▶ One size infinity fits all?

We can count the even numbers, the square numbers, the rational numbers. This suggests that there is only one infinity: the infinity of the natural numbers is the same size as the infinity of the even numbers, the same size as the infinity of the rational numbers and so on. The work of Georg Cantor showed that it was not that simple. It turns out that there are some infinities that are larger than others.

Cantor took this basic idea of counting as putting things into one-to-one correspondence to develop the branch of mathematics known as set theory, which is the idea of taking collections of, for example, numbers, and asking questions about the cardinality of the collections. The leap of imagination that Cantor took was to think about the cardinality of infinite sets. Up to that point, the prevailing view of infinity was encapsulated in Gauss's claim that infinity was 'merely a way of speaking'. Infinity was not treated as a mathematical 'object'. Through his exploration of the cardinality of sets, Cantor brought infinity into being as an object that could be examined in its own right. And in doing so he established that there is more than one infinity.

As we saw above, some infinite sets are, in Cantor's terms, countable, i.e. can be put into one-to-one correspondence with the natural numbers. These countable sets all have the same cardinality. Although there was a symbol for infinity – ∞ – introduced by John Wallis in the 1700s, Cantor introduced a new symbol for the cardinality of these countable infinite sets – aleph naught (or aleph null), the first letter of the Hebrew alphabet with a subscript 0. Cantor theorized (correctly as it has come to be accepted) that this is one of a number of different infinities – **transfinite numbers**, in his term – and the smallest of such numbers. Cantor's genius was to show that the infinite set of irrational numbers is not countable, that its cardinality is an infinity greater than that of the natural or rational numbers. Not surprisingly, this did not go down well with his contemporaries, although it is now accepted in mainstream mathematics.

## ▶ To infinity and beyond

To explore how Cantor showed that the irrationals are not countable, we return to proof by contradiction.

Cantor's conjecture was that the irrational numbers are not countable. Recall that proof by contradiction starts from the opposite premise: the assumption that the irrational numbers *are* countable. From this assumption we look at the logical conclusions we can draw, until reaching a logical contradiction, an absurdity. Here goes.

If the real numbers are countable, that means we could, in theory, list out all the natural numbers and match them as before, in one-to-one correspondence, with the natural numbers. It helps to list the natural numbers in order, but if we can count all of the irrationals, then it does not matter what order we put them in so long as we list them all. Let's list and count a few:

1 0.4568972 ...

2 0.7503886 ...

3 0.1243962 ...

4 0.9756441 ...

5 0.1818232 ...

Irrational numbers when expressed as decimals have digits in the decimal places that never terminate or repeat – these ellipses indicate that each of these numbers has an infinite number of such decimal places.

Cantor's first step was to come up with a rule for using such a list of irrational numbers to create another irrational number by taking the digit in the first decimal place of the first number, the digit in the second decimal place of the second number, the digit in the third decimal place in the third number, and so on. From my list, this new number would start:

0.45462 ...

This new number would, of course, be somewhere in the overall list. Cantor's brilliant trick was to use this new number to create another irrational number that cannot be in the list.

To do this, all we have to do is simply(!) change every digit in this number. It does not matter what new digit each existing digit is changed to; it just has to be changed. So let me turn this number into a new one by adding one to each digit.

0.56573 ...

This new number definitely is different from the first number in my list, because the digit in the first decimal place is different. It is also different from the second number in the list, because the digit in the second decimal place is different, and, by the same logic, it is different from the third, 20th or 153rd number in the list. No matter how far down the list we go (and it is an infinite list), the number just created is different from every other number in the list. We have to conclude that this number is not in the list. But we started off with the assumption that all the irrational numbers *were* in the list (i.e. that they would all be there if they were countable). The only conclusion we can reach is that our original assumption was incorrect – the real numbers are not countable. The infinity of the irrational is larger than the infinity of the natural numbers.

During Cantor's lifetime, this work so challenged prevailing views that it was widely criticized. Now is it accepted as a fundamental theory and the mathematical community agrees with the mathematician David Hilbert's description of Cantor's work as 'the finest product of mathematical genius'. Sadly, Cantor suffered recurrent bouts of severe depression resulting in frequent hospitalization, possibly exacerbated by the hostile reaction to his ideas.

Once accepted into the canon of mathematical knowledge, Cantor's transfinite numbers have encouraged mathematicians to explore other ways of visualizing the number system. One such image is **Ford circles**, developed by the mathematician Lester Ford, Snr. The imagery is a little like the foam in a bubble bath – a collection of bubbles all touching each other and not intersecting. Smaller bubbles fill any space between bubbles. Ford circles provide mathematicians with a powerful visualization of different sizes of infinity – and perhaps to follow in the footsteps of Archimedes and get inspiration while having a soak in the bath! Which, as we have reached the end of our journey, you may now consider treating yourself to.

# Five books on the philosophy of mathematics

100 IDEAS

1  *The Mathematical Experience* (1981), Philip J. Davis and Reuben Hersh. A classic text describing what mathematics is about and what mathematicians do.

2  *A Mathematician's Apology* (1940), G. H. Hardy. An essay concerning the aesthetics of mathematics, and one of the best insights into the mind of a mathematician.

3  *Gödel, Escher, Bach: An Eternal Golden Braid* (1979), Douglas R. Hofstadter. An examination of the nature of human thought processes which won the Pulitzer Prize for general non-fiction in 1980.

4  *Mathematics in Western Culture* (1953), Morris Kline. Argues that mathermatics has been a major cultural force in western civilization, underpinning physical and social sciences, philosophy and religion, and influencing art, architecture, music and literature.

5  *Proofs and Refutations: The Logic of Mathematical Discovery* (1963–64), Imre Lakatos. A famous dialogue, first published in the *British Journal for the Philosophy of Science*, which considers the methodology, philosophy and history of mathematics.

# Five biographies of mathematical ideas

6  *The Nothing That Is: A Natural History of Zero* (1999), Robert Kaplan. Traces the development of the concept of 'nothing' and its mathematical representation, zero.

7  *Zero: The Biography of a Dangerous Idea* (2000), Charles Seife. Also traces the history of zero, and its impacts on eastern and western cultures, religion and science.

8  *Fermat's Last Theorem* (1997), Simon Singh. An accessible account of the efforts to solve the 350-year-old problem and how Andrew Wiles finally succeeded in the 1990s.

9  *The Music of the Primes: Why an Unsolved Problem in Mathematics Matters* (2003), Marcus du Sautoy. Relates the struggles of mathematicians with the most difficult outstanding problem in mathematics, the Riemann hypothesis.

10  *A Brief History of Infinity: The Quest to Think the Unthinkable* (2003), Brian Clegg. A popular survey of the paradoxes that infinity poses and how mathematicians and philosophers have grappled with the concept.

# Five classic (and accessible) mathematical problems

11  Fermat's last theorem. This states that no three positive integers, $a$, $b$ and $c$, can satisfy the equation $a^n + b^n = c^n$; first conjectured by Pierre de Fermat in 1637; first successfully proved by Andrew Wiles in 1994.

12  The four colour theorem. This states that given any separation of a plane into contiguous regions, called a map, the regions can be coloured using at most four

colours so that no two adjacent regions have the same colour; first conjectured in 1852; first successfully proved by Kenneth Appel and Wolfgang Haken in 1976.

**13** The seven bridges of Königsberg. The problem was to find a walk through the Prussian city (now Kaliningrad, Russia) that would cross once and only once each of the seven bridges connecting the two banks of the Pregel river and the two islands in the river with each other; Leonhard Euler's proof in 1735 that the problem has no solution laid the foundations of graph theory.

**14** Goldbach's conjecture. This states that every even integer greater than 2 can be expressed as the sum of two primes; first conjectured by Christian Goldbach in 1742; the conjecture holds through to $4 \times 10^{18}$ and is generally assumed to be true but remains unproven.

**15** Squaring the circle. This problem is the challenge of constructing a square with the same area as a given circle using only a finite number of steps with compass and straight-edge; first posed by ancient mathematicians; proven impossible in 1882 by Ferdinand von Lindemann.

# Five books on how we learn mathematics (or don't)

**16** *The Number Sense: How the Mind Creates Mathematics* (2011), Stanislas Dehaene. Argues that a sense of quantity is hard-wired into us.

**17** *Where Mathematics Comes From: How the Embodied Mind Brings Mathematics into Being* (2000), George Lakoff, Rafael E Núñez. Provides a cogent argument that all mathematics, even the most abstract, can be traced back to bodily experiences and sensations, which act as metaphors for more general ideas.

**18** *Innumeracy: Mathematical Illiteracy and its Consequences* (1990), John Allen Paulos. An amusing and thought-provoking look at the impact of lack of mathematical awareness on everyday life.

**19** *A Mathematician's Lament: How School Cheats Us Out of Our Most Fascinating and Imaginative Art Form* (2010), Paul Lockhart. A brief account of the false impression that schooling gives of mathematics as a discipline.

**20** *Thinking Mathematically* (2010), John Mason, Leone Burton, Kaye Stacey. Lots of accessible mathematical activities to work on.

# Ten key mathematicians

**21** **Pythagoras** (*c.*570–*c.*495 BCE). A Greek philosopher, mystic and mathematician famous to school pupils because of the Pythagorean theorem. Although many sources now question whether Pythagoras constructed the proof bearing his name, he is widely regarded as the founding father of modern mathematics.

**22** **Euclid** (active 300 BCE). A Greek mathematician teaching in Alexandria, Euclid is generally known as the 'father of geometry' through his 13-volume *Elements*. He is credited with introducing rigorous, logical proof for theorems, an approach that has shaped mathematical activity since his time.

**23** **Leonardo Bonacci** (*c.*1170–*c.*1250), popularly known as Fibonacci. One of the most famous mathematicians of the Middle Ages, Fibonacci was influential in popularizing use of the modern Hindu–Arabic numeral system in Europe and the 'Fibonacci sequence' of numbers (although Indian mathematicians had known this sequence since *c.*200 BC).

**24** **René Descartes** (1596–1650). Famous for Cartesian coordinates, Descartes is revered for his work on analytical geometry, bringing algebra and geometry together.

**25** **Sir Isaac Newton** (1642–1727). Although popularly known for his work on gravity, Newton made a major contribution to mathematics through his work on calculus. There is debate over whether Newton or Leibniz was the first to develop calculus: what is clear is that both made considerable contributions.

**26** **Leonhard Euler** (1707–83). Of Euler's multitude of contributions to mathematics, one of the most significant was the introduction of much mathematical notation that is still in use today. His work included calculus, topology, number theory, analysis and graph theory.

**27** **Carl Friedrich Gauss** (1777–1855). A child prodigy, Gauss, the 'prince of mathematics', had established a mathematical reputation by the age of 21. He made major contributions in many areas of mathematics, most notably in number theory, proving the fundamental theorem of algebra. Prolific up to his death, his work has had a remarkable influence on mathematics.

**28** **Bernhard Riemann** (1826–66). Riemann was one of the most prominent mathematicians of the 19th century, as witnessed by the number of theorems attributed to him, including Riemannian geometry, Riemannian surfaces, the Riemann integral, and in particular the Riemann hypothesis, one of the great unsolved problems of mathematics, regarding the patterning of prime numbers.

**29** **Georg Cantor** (1845–1918). Famous for his work on infinity, Cantor had few contemporaries with whom he could discuss this groundbreaking work; in his lifetime his ideas were met with strong resistance, but they are now regarded as some of the most important in all of mathematics.

**30** David Hilbert (1862–1943). A revered mathematician in his own right, Hilbert is popularly remembered for presenting at the International Congress of Mathematicians in Paris in 1900 a list of 23 unsolved problems (ten in his address and the others in the printed proceedings) that he considered as important for contemporary mathematics – probably the most considered compilation of open problems produced by an individual mathematician.

# Five women mathematicians

**31** Hypatia (c.350 or 370– c.415 AD). A Greek astronomer and mathematician working in Alexandria, Hypatia was possibly not the first female mathematician, but her work has been passed down to later mathematicians. Her studies of conic sections – parabolas, hyperbolas and ellipses – provided the foundation for later work by Descartes, Newton and Liebniz.

**32** Ada, Countess of Lovelace (1815–52). The daughter of Lord Byron (whom she never knew), Lovelace struck up a correspondence with the mathematician Charles Babbage. When translating an Italian mathematician's memoir for him, Lovelace realised that a calculating maching could be developed to run off punched cards in the way that weaving machines did, establishing her as one of the earliest pioneers of computer programming; the ADA computer programming language was named after her.

**33** Sofia Kovalevskaya (1850–91). Kovalevskaya left Russia, where women could not attend lectures, and was eventually appointed lecturer in mathematics at the University of Stockholm and later became the first woman in that part of Europe to receive a full professorship. Her work was further recognized by the award of the Prix Bordin from the French Academy of Sciences (1888) and a prize from the Swedish Academy of Sciences (1889).

**34** Emmy Noether (1882–1935). Described by Einstein as 'the most significant creative mathematical genius thus far produced since the higher education of women began', Noether struggled to achieve recognition in her native Germany, as women could not then hold a full academic post. Even her title of 'unofficial associate professor' at the University of Göttingen was removed in 1933 because she was Jewish. After emigrating to the USA, her work there provided mathematical foundations for Einstein's general theory of relativity.

**35** Maryam Mirzakhani (1977– ). In 2014 Maryam Mirzakhani became the first woman as well as the first Iranian to be awarded a Fields Medal (officially the International Medal for Outstanding Discoveries in Mathematics), a prize awarded to mathematicians not over 40 years of age; 55 medals have been awarded since 1936. Mirzakhani, who works in the USA, was recognized for her work on complex geometry.

# Five biographies of mathematicians

**36** *Euler: The Master of Us All* (1999), William Dunham. Explores the huge scope of mathematical areas explored and developed by Leonard Euler.

**37** *Alan Turing: The Enigma* (1983), Andrew Hodges. Turing (1912–54) masterminded the cracking of Germany's Enigma code during World War II and made an outstanding contribution to computer science.

**38** *The Man Who Loved Only Numbers: The Story of Paul Erdös and the Search for Mathematical Truth* (1998), Paul Hoffman. Paul Erdös (1913–96), a key figure in pure mathematics, had an eccentric lifestyle devoid of virtually everything except numbers.

**39** *The Man Who Knew Infinity* (1991), Robert Kanigel. A biography of the self-taught Indian mathematical prodigy Srinivasa Ramanujan (1887–1920).

**40** *A Beautiful Mind* (1998), Sylvia Nasar. A biography of the American mathematician John F. Nash, Jnr (1928– ), who overcame mental illness to share a Nobel Prize in 1994 for pioneering work on game theory.

# Five books on women in mathematics

**41** *Women in Mathematics: The Addition of Difference* (1997), Claudia Henrion. Studies how gender and race affect women's participation in mathematics.

**42** *Hypatia's Heritage: A History of Women in Science from Antiquity through the Nineteenth Century* (1986), Margaret Alic.

**43** *Women Becoming Mathematicians: Creating a Professional Identity in Post-World War II America* (2000), Margaret Murray.

**44** *The Bride of Science: Romance, Reason and Byron's Daughter* (1999), Benjamin Woolley. Biography of Ada, Countess of Lovelace.

**45** *A to Z of Women in Science and Math* (2007), Lisa Yount.

# Ten popularisers of mathematics

**46** Alex Bellos. British journalist turned author and broadcaster of works about football, Brazil and mathematics.

**47** **Keith Devlin**. British-born and US-domiciled mathematician, academic and popular science writer and broadcaster.

**48** **Rob Eastaway**. Popular writer on mathematics and sport, and director of Maths Inspiration, a national programme of maths lectures for teenagers.

**49** **Jordan Ellenberg**. An American child mathematics prodigy, now an academic and popularizer of mathematics through books, articles and the internet.

**50** **Martin Gardner**. Prolific American author of popular mathematics and science journalism who, via his 'Mathematical Games' column in *Scientific American* from 1956 to 1981, did much to nurture interest in recreational mathematics in the USA.

**51** **Sir Timothy Gowers**. British mathematician and academic who won the Fields Medal (mathematical equivalent of a Nobel Prize) in 1998. He popularizes mathematical matters through several non-fiction works and his blog, where he has encouraged collaborative solving of mathematical problems.

**52** **Matt Parker**. British mathematician and academic known as the Stand-up Mathematician because of his stand-up performances. He founded the Think Maths team to present maths talks and workshops in schools.

**53** **Simon Singh**. British BBC television producer turned author, journalist and TV presenter of mathematical and scientific subjects, including cryptography.

**54** **Diana Taimina**. Latvian mathematician and academic working in the USA, known for using crocheting to demonstrate and explore mathematical concepts in her presentations to the wider public.

**55** Margaret Wertheim. Australian-born science writer and journalist who, with her twin sister, founded the Institute for Figuring to promote public understanding of the aesthetic dimensions of science and mathematics; one project, the 'Hyperbolic Crochet Coral Reef', is one of the world's largest arts and science community projects.

# Ten quotations about mathematics

**56** 'The essence of mathematics is its freedom.' Georg Cantor

**57** '[T]he different branches of Arithmetic – Ambition, Distraction, Uglification, and Derision.' Lewis Carroll

**58** 'Pure mathematics is, in its way, the poetry of logical ideas.' Albert Einstein

**59** 'A mathematician is a device for turning coffee into theorems.' Paul Erdös

**60** 'The laws of nature are but the mathematical thoughts of God.' Euclid

**61** 'Mathematics knows no races or geographic boundaries; for mathematics, the cultural world is one country.' David Hilbert

**62** 'Music is the pleasure the human mind experiences from counting without being aware that it is counting.' Gottfried Leibniz

**63** 'We could use up two Eternities in learning all that is to be learned about our own world and the thousands of nations that have arisen and flourished and vanished from it. Mathematics alone would occupy me eight million years.' Mark Twain

**64** 'If people do not believe that mathematics is simple, it is only because they do not realize how complicated life is.' John Louis von Neumann

**65** 'A man has one hundred dollars and you leave him with two dollars. That's subtraction.' Mae West

# Five mathematical museums or galleries

**66** National Museum of Mathematics, New York, USA; http://momath.org

**67** Haus der Mathematik (Mathematics Museum), Vienna, Austria; www.hausdermathematik.at/

**68** Das Mathematikum, Giessen, Germany; www.mathematikum.de

**69** The Institute for Figuring, Los Angeles, USA; www.theiff.org

**70** The Virtual Math Museum; http://virtualmathmuseum.org/

# Five ancient mathematical buildings

**71** The Pantheon, Rome, Italy. Has the mathematical perfection of a dome on a cubic base.

**72** The Parthenon, Athens, Greece. Its proportions represent the Golden Ratio.

**73** The Great Pyramid of Giza, Cairo, Egypt. A structure full of mathematical connections, such as that the pyramid's perimeter divided by twice its height is equal to pi.

**74** The Alhambra Palace, Granada, Spain. Contains tiling tessellations with examples of the 17 possible underlying mathematical structures.

**75** The Taj Mahal, Agra, India. A building that is bursting with symmetries.

# Five modern mathematical buildings

**76** The Eden Project, Cornwall, England. A geodesic dome made up of pentagons and hexagons.

**77** The Guggenhiem Museum, Bilbao, Spain. A glorious example of mathematics and computer-aided design coming together.

**78** The AAMI Park Stadium, Melbourne, Australia. Remarkable for the complex geometry of the roof.

**79** The US Air Force Academy Cadet Chapel, Colorado Springs, USA. A groundbreaking design of 17 spires build on a tetrahedral frame.

**80** The Philips Pavilion, Brussels, Belgium. A building comprising symmetric hyperbolic paraboloids.

# Five great artworks based on mathematics

**81** The Flagellation of Christ by Piero della Francesca. A prime example of the early use of linear perspective.

**82** Vitruvian Man by Leonardo da Vinci. Sets out significant points of the proportion of the body and the architectural representation of the body.

**83** Crucifixion (Corpus Hypercubus) by Salvador Dalí. Employs a net of a hypercube as the cross.

**84** Circle Limit III and Circle Limit IV prints by Maurice Escher. Close to illustrating hyperbolic geometry.

**85** Composition with Yellow, Blue, and Red by Piet Mondrian. Typical composition using only vertical and horizontal lines at 90-degree angles.

# Five novels with a mathematical theme

**86** *Uncle Petros and Goldbach's Conjecture* (2000), Apostolos Doxiadis. A novel of mathematical obsession that explores the spirit of mathematical research.

**87** *The Solitude of Prime Numbers* (2009), Paulo Giordano, trans. Shaun Whiteside. An examination of loneliness and love through the lives of two emotionally damaged people.

**88** *The Curious Incident of the Dog in the Night-time* (2003), Mark Haddon. A murder mystery investigated by a 15-year-old boy with a form of autism who understands maths but not people.

**89** *The Oxford Murders* (2005), Guillermo Martínez, trans. Sonia Soto. A mathematician has to work out the key to coded messages before a serial killer strikes again.

**90** *The Housekeeper and the Professor* (2009), Yoko Ogawa, trans. Stephen Snyder. Explores how relationships are formed in a story about how a maths professor with only 80 minutes of short-term memory relates to his housekeeper and her son.

# Five films with a mathematical theme

**91** *A Beautiful Mind* (2001); director, Ron Howard; starring Russell Crowe, Ed Harris and Jennifer Connolly. A screen version of Sonia Nasar's biography of John Nash.

**92** *Cube* (1997); director, Vicenzo Natali; starrring Maurice Dean Wint, David Hewlett, Nicole de Boer, Nicky Guadagni, Andrew Miller. A group trapped inside a booby-trapped cube use mathematics to escape.

**93** *Good Will Hunting* (1997); director, Gus Van Sant; starring Matt Damon, Robin Williams, Ben Affleck. A janitor with a gift for mathematics needs help from a psychologist to find direction in his life.

**94** *Pi* (1998); director, Darren Aronofsky; starring Sean Gullette, Mark Margolis, Ben Shenkman. A paranoid mathematician searches for a key number that will unlock the universal patterns found in nature.

**95** *Sneakers* (1992); director, Phil Alden Robinson; starring Robert Redford, Dan Aykroyd, Sidney Poitier, Ben Kingsley, Mary McDonnell, River Phoenix, David Strathairn. A caper movie about security analysts blackmailed into stealing a chip capable of decoding all existing encryption systems.

# Five mathematical animations

**96** Dance Squared at www.youtube.com/watch?v=yXL4DP_3dJI (accessed 15 September 2014).

**97** Rythmetic at www.youtube.com/watch?v=xWRRAw6xzos (accessed 15 September 2014).

**98** Donald in Mathmagic Land at www.youtube.com/watch?v=nav0kVa66xk (accessed 15 September 2014).

**99** The Dot and the Line at www.youtube.com/watch?v=OmSbdvzbOzY (accessed 15 September 2014).

**100** Beauty of Mathematics at http://vimeo.com/77330591 (accessed 15 September 2014).

# Selected bibliography

Bellos, A., *Alex's Adventures in Numberland* (London; New York: Bloomsbury Paperbacks, 2011).

Clegg, B., *Brief History of Infinity: The Quest to Think the Unthinkable* (London: Robinson Publishing, 2003).

Crilly, T., *50 Mathematical Ideas You Really Need to Know* (London: Quercus, 1st edn, 2008).

Cuoco, A., Goldenberg, E. P. and Mark, J., 'Habits of mind: An organizing principle for a mathematics curriculum', *Journal of Mathematical Behavior*, 15, (1996) 375–402.

Davis, P. J. and Hersh, R., *The Mathematical Experience* (Boston: Houghton Mifflin, 1981).

Devlin, K., *Language of Mathematics* (New York: Palgrave Macmillan, reissued edn, 2003).

Ellenberg, J., *How Not to be Wrong: The Hidden Maths of Everyday Life* (London: Allen Lane, 2014).

Elwes, R., *Maths in 100 Key Breakthroughs* (London: Quercus, 2013).

Flegg, G., *Numbers: Their History and Meaning* (Harmondsworth: Pelican, 1984).

Frenkel, E., *Love and Math* (New York: Basic Books, 2013).

Gowers, T., *Mathematics: A Very Short Introduction* (Oxford; New York: Oxford Paperbacks, 2002).

Houston, D. K., *How to Think Like a Mathematician: A Companion to Undergraduate Mathematics* (New York: Cambridge University Press, 2009).

Kaplan, R. and Kaplan, E., *Out of the Labyrinth: Setting Mathematics Free* (Oxford; New York: Oxford University Press, 2007).

Körner, T. W., *The Pleasures of Counting* (Cambridge, UK; New York: Cambridge University Press, 1996).

Mason, J., Burton, L. and Stacey, K., *Thinking Mathematically* (Harlow, UK: Prentice Hall, 2nd edn, 2010).

Peterson, I., *The Mathematical Tourist: Snapshots of Modern Mathematics* (New York: Freeman and Co., 1988)

Pickover, C. A., *The Math Book* (New York: Sterling, 2012).

Pitici, M. (ed), *The Best Writing on Mathematics 2013* (Princeton, NJ: Princeton University Press, 2014).

Polkinghorne, J., *Meaning in Mathematics* (Oxford; New York: Oxford University Press, 2011).

Polya, G., *How to Solve it: A New Aspect of Mathematical Method* (London: Penguin, 1990).

Sautoy, M. D., *The Music of the Primes: Why an Unsolved Problem in Mathematics Matters* (London: Harper Perennial, 2004).

Sautoy, M. D., *Finding Moonshine: A Mathematician's Journey Through Symmetry* (London: Harper Perennial, 2009).

Stewart, I., *From Here to Infinity* (Oxford; New York: Oxford Paperbacks, 3rd edn, 1996).

Wells, D., *The Penguin Dictionary of Curious and Interesting Numbers* (London; New York: Penguin, revised edn, 1997).

# Index

AAMI Park Stadium 144
abstract 110–11
Ada, Countess of Lovelace 138, 140
addition 12
Alhambra Palace 99, 144
Argand diagram 50
associative 58–9

barber paradox 120–1
Bellos, Alex 140
Bonacci, Leonardo 136

Cantor, Georg 125, 127–31, 137, 142
cardinal aspect 10–11
Carroll, Lewis 142
Cartesian coordinates 103–8
Cauchy, Augustin-Louis 64–5
Cayley tables 94–7
*Circle Limit III* and *Circle Limit IV* 145
coastline paradox 115–16
commutative law 56–9
complex numbers 50
composite numbers 60
*Composition with Yellow, Blue, and Red* 145
conjectures 68, 82–3
counting 10–20
  infinity 122–31
*Crucifixion* 144

deduction, proof by 71–7
Descartes, René 103–5, 137
Devlin, Keith 141
disproving 82–3
distance between two points 105–8
distributive law 59–60
division 12, 37–8
  fractions 42–4

Eastaway, Rob 141
Eden Project 144
Egyptian fractions 39–40
Einstein, Albert 112, 139, 142
Ellenberg, Jordan 141
embodied mathematics 5–6
Erdös, Paul 139, 142
Euclid 29–32, 136, 142
Euclidean geometry 28–32
Euler, Leonard 135, 137, 139

Federov, E. S. 99
Fermat, Pierre de 108, 134
Fermat's last theorem 134
Fibonacci sequence 136
*Flagellation of Christ* 144
Ford, Snr, Lester 131
Ford circles 131
four colour theorem 134–5
fractals 116–17
fractions 38–45
fuzzy logic 121–2

Galois, Évariste 98
Gardner, Martin 141
Gauss, Carl Friedrich 50, 74, 127, 137
generalizing 36–50
geometry 22–34
Gödel, Kurt 121
Goldbach's conjecture 135
Gowers, Sir Timothy 141
Great Pyramid of Giza 143
Griess, Robert 100
groups 94–100
Guggenheim Museum 144

handshakes problem
  proof by deduction 71–7

proof by induction 77–9
  structure 80–2
Hilbert, David 130, 138, 142
Hypatia 138
hyperbolic geometry 33–4

identity element 12–13
identity transformations 89
imaginary numbers 49–50
'in general', as a phrase 57
incompleteness theory 121
induction, proof by 77–9
infinity, counting 122–31
integers 10
invert-and-multiply rule 42–4
irrational numbers 45–50

Koch snowflake 113–14
Kovalevskaya, Sofia 3, 138

Leibniz, Gottfried 137, 142

Mandelbrot, Benoit 116–17
Mason, John 55
mathematics, as invention or
  discovery 7–8
Mirzakhani, Maryam 139
modelling, Cartesian
  coordinates 103–8
modelling cycle 102–3
monster group 100
multiplication 12

natural numbers 10–12
negative numbers 16–20
Neumann, John Louis von 143
Newton, Sir Isaac 137
Noether, Emmy 139
non-Euclidean geometry
  32–4
number lines 40–2

ordinal aspect 10–11

Pantheon 143
parallels 33–4
Parker, Matt 141
Parthenon 143
pattern sniffing 52–66, 80–1
Philips Pavilion 144
place value 13–14
prime numbers 60–2
proof
  by deduction 71–7
  by induction 77–9
  by looking at structure 80–2
  conjectures 68–71, 82–3
  disproving 82–3
Pythagoras' theorem 22–8, 136
Pythagoreans 39, 45–7

rational numbers 39
real numbers 49
Richardson, Lewis 115–16
Riemann, Bernhard 62, 137
rules of arithmetic 54–60
Russell, Bertrand 120–1

school mathematics 4–5
self-similarity 116–17
set theory 121
seven bridges of Königsberg 135
Singh, Simon 141
squares 22–6
squaring the circle 135
string theory 112–13
structure, proof by looking at
  80–2
subtraction 12
symmetry 86–93

Taimina, Diana 33–4, 141
Taj Mahal 144

transfinite numbers 128, 131
transformations 87–93
triangles 25–32
triangular numbers 53
Twain, Mark 142

ultraparallels 33–4
unitary fractions 40
US Air Force Academy Cadet
    Chapel 144

visualizing 22–34
*Vitruvian Man* 144

Wallis, John 17–18
wallpaper groups 98–9
Wertheim, Margaret 142
West, Mae 143

Zadeh, Lotfi 122
zero 12–16

ALL THAT MATTERS: MATHEMATICS

*All That Matters* books are written by the world's leading experts, to introduce the most exciting and relevant areas of an important topic to students and general readers.

From Bioethics to Muhammad and Philosophy to Sustainability, the *All That Matters* series covers the most controversial and engaging topics from science, philosophy, history, religion and other fields. The authors are world-class academics or top public intellectuals, on a mission to bring the most interesting and challenging areas of their subject to new readers.

Each book contains a unique '100 Ideas' section, giving inspiration to readers whose interest has been piqued and who want to explore the subject further. Find out more, at:

www.allthatmattersbooks.com
Facebook/allthatmattersbooks
Twitter@All_That_Matters